Praise for

CANADIAN COPYRIGHT

"Sophisticated and thorough in its approach, using up-to-date cases and examples, *Canadian Copyright: A Citizen's Guide* presents an accessible and engaging explanation of intellectual property law. Murray and Trosow capture the complexities of the Canadian tradition, drawing on recent court decisions by Madame Justice Rosalie Abella and reaching back to the wisdom of Northrop Frye. This book is ultimately empowering, with knowledge that you can use when it comes to intellectual property."

— JOHN WILLINSKY, Professor of Publishing Studies, Simon Fraser University, and Khosla Family Professor, Stanford University

"As an artist who is actively seeking to broaden my reach, I frequently adopt new ways to connect with my audience online. *Canadian Copyright: A Citizen's Guide* helped me understand how copyright works in Canada and how I can use it to protect my work and message as an artist."

— SONNY ASSU, interdisciplinary artist, Montreal

"Murray and Trosow's book is essential reading for anyone anywhere who wants to understand how—encouraged by Canadian civil society—Parliament and the courts have taken a lead role in the worldwide struggle for balanced copyright law. *Canadian Copyright* points out the struggles that lie ahead, in particular with regard to assuring that copyrighted content is fully available for educational use. Accurate without being hyper-technical, wonderfully readable, and with a consistent emphasis on how choices about copyright law affect real-life patterns of cultural production and consumption, this volume is truly a model of its kind."

— PETER JASZI, Professor, American University Law School

"In an area of public debate often marked with frustrating polemic and oversimplification, *Canadian Copyright: A Citizen's Guide* provides a nuanced, articulate, accessible, and, perhaps most importantly, uniquely Canadian perspective. This book is essential reading for students not only of copyright, but also of Canadian culture."

— KEITH SERRY, President, Clinique Juridique des Artistes de Montréal/The Montreal Artists Legal Clinic Co-founder, Canadian Music Creator's Coalition

"Trosow and Murray offer an insightful account of copyright's relevance to all of us— citizens, academics, innovators, and creators of all stripes. Updated to reflect the latest word on Canadian copyright law—from both Parliament and the Supreme Court—*Canadian Copyright: A Citizen's Guide* offers a much-needed account of how copyright affects all of us."

— DAVID FEWER, Director of the Samuelson-Glushko Canadian Internet Policy & Public Interest Clinic

CANADIAN COPYRIGHT

A CITIZEN'S GUIDE

—— SECOND EDITION ——

LAURA J. MURRAY & SAMUEL E. TROSOW

BETWEEN THE LINES   TORONTO

Canadian Copyright: A Citizen's Guide, Second edition
© 2013 Laura J. Murray and Samuel E. Trosow

First published in 2013 by
Between the Lines
401 Richmond Street West
Studio 277
Toronto, Ontario M5V 3A8
Canada
1-800-718-7201
www.btlbooks.com

Every reasonable effort has been made to identify copyright holders. Between the Lines
would be pleased to have any errors or omissions brought to its attention.

Library and Archives Canada Cataloguing in Publication

Murray, Laura Jane, 1965–
     Canadian copyright : a citizen's guide / Laura J. Murray and Samuel E. Trosow ;
illustrator, Jane Burkowski.—2nd ed.

Includes bibliographical references and index.
Issued also in electronic formats.

ISBN 978-1-77113-013-4

1. Copyright—Canada—Popular works.   2. Copyright—Canada—Cases—Popular works.
I. Trosow, Samuel E   II. Title.

KE2799.2.M87   2013   346.7104>82   C2012-907736-4
KF2995.M87   2013

Text design by Gordon Robertson
Front cover photo © Dan Kosmayer / Shutterstock.com
Printed in Canada

RECYCLED
Paper made from
recycled material
FSC® C103567

Between the Lines gratefully acknowledges assistance for its publishing activities from
the Canada Council for the Arts, the Ontario Arts Council, the Government of Ontario
through the Ontario Book Publishers Tax Credit program and through the Ontario
Book Initiative, and the Government of Canada through the Canada Book Fund.

Canada Council
for the Arts
Conseil des Arts
du Canada
Canadä
ONTARIO ARTS COUNCIL
CONSEIL DES ARTS DE L'ONTARIO
50 YEARS OF ONTARIO GOVERNMENT SUPPORT OF THE ARTS
50 ANS DE SOUTIEN DU GOUVERNEMENT DE L'ONTARIO AUX ARTS

Contents

List of Tables vii

Introduction ix

PART I: IDEAS

1. Copyright's Rationales 3
2. Copyright's Histories 16

PART II: LAW

3. Copyright's Scope 35
4. Owners' Rights 54
5. Users' Rights 71
6. Collectives and the Copyright Board 87
7. Determining Ownership 93
8. Enforcement of Owners' Rights 101

PART III: PRACTICE

9. Music 121
10. Digital Media 134
11. Film, Video, and Photography 149

12. Visual Arts   158

13. Craft and Design   165

14. Journalism   172

15. Education   182

16. Libraries, Archives, and Museums   199

PART IV: CONTEXTS

17. Copyright's Counterparts   217

18. Copyright's Future   234

Notes   238

Legal Citations and Cases   262

Bibliography   265

Index   272

Tables

1. Comparison of Pure Public Goods and Pure
   Private Goods   13
2. Recent Copyright Reform Bills   28
3. Major Types of Intellectual Property   36
4. Works and Other Subject Matter Covered
   by Copyright   38
5. Duration of Copyright Term in Special Cases   50
6. Duration of Copyright for Unpublished
   or Posthumously Published Works   52
7. Rights in Works and Other Subject Matter   55
8. Exceptions to Owners' Rights of General
   Application   84
9. Rules Governing First Ownership   95
10. Civil and Criminal Law Cases   102
11. Domains of User-Generated Content   147
12. Exceptions Applicable to Educational Institutions   189

Introduction

These days copyright has become part of just about everyone's life. That's why you are reading this book. Whether you are a parent, artist, business person, blogger, teacher, student, or music fan, questions about copyright law have popped into your head or landed in your lap. You may want to stop people from using your screenplay or photograph without your permission. You may wonder whether you should read all that legalese on a software licence or a publishing contract, and whether you'd understand it if you did. You may want to know if it's okay to capture an image from somebody else's website and post it on your own. You may wish you knew how to argue with a boss, a teacher, or a lawyer who says, "You can't do that."

In this book, we seek to help you out with these practical questions. But we admit right off the bat that this is not a "dummies' guide." We'll take you through some history and philosophical underpinnings on the way to the answers. Copyright law, like all law, is not like a series of switches. It's a human creation. And it is still very much still a work in progress. Over the years, and in different countries, it has become diversified as it is adapted to many new situations and technologies.

Most people see copyright law, along with law in general, as static: some things are illegal, some things are legal, and the judge will tell us which is which. If it isn't static, many people think it ought to be: that with the right tools we can immobilize copyright law and make it more certain. But in fact all law is always developing in a complex and fitful way—through changing legislation, through legal precedents from case law, and through the practice and beliefs of ordinary citizens. Law is not a thing, but a process based on a set of social relationships. For many people this aliveness of the law produces confusion, but in copyright, as in other areas of law, we think it also produces opportunities. If ordinary people educate themselves about the history and various incarnations of copyright around the world, they can glimpse principles, costs, and possibilities often masked by the misleading self-evidence of the here and now. Widespread knowledge of the law can enable people to make more effective use of it—in our terms, to practise fair copyright.

A sense of popular empowerment and responsibility is just as important now as it was in 2007 when the first edition of this book came out. At that time, Canada was in the midst of a major debate over what direction legislative reform should take. In fact, we were worried that our book would become out of date within months of its publication! But the legislative reform did not happen until 2012, in the form of Bill C-11. That was a big year for copyright in Canada: only shortly after C-11 passed, and even before it was enacted into law, the Supreme Court delivered five copyright cases that have an important bearing on users' rights and technological neutrality issues. At around the same time, a large number of universities decided not to renew their licences with the major educational copyright collective, Access Copyright. So, even though Access Copyright has continued with its strategy of lawsuits and tariff applications, the legal landscape is very different now than it was in 2007. It is more certain in some ways, but the challenge is for Canadians to take up and inhabit the new environment.

Parliament and the courts haven't been the only movers of change when it comes to copyright. In fact, we would say that it is primarily ordinary people who have changed the tone and results of copyright discussions and practice in the years since 2007. At that time, we described a climate of fear and threat, in which content providers seemed to think of their customers as pirates, and libraries and educational institutions were cowed into a very narrow view of users' rights. But Canadians from a wide range of positions and professions started to really pay attention over the past few years. Copyright bills introduced in Parliament in 2005, 2008, and 2010 were highly controversial. All three were abandoned, for procedural rather than political reasons, but they generated a lot of heat before they died. Consumers struck back against provisions they considered unfair, via Facebook, campaign trail showdowns, and the like. Meanwhile, the cultural industries and cultural workers stood up for owners' rights, though many artists and musicians acknowledged the importance of users' rights as well. There have been some pretty dramatic moments in Canadian copyright over the past few years. A very technical part of the law once of interest only to publishers and lawyers became, at least for a while, a topic of discussion in bars and coffee shops all over the country.

We hope those discussions continue and, as the dust settles, become less polarized. We have always insisted that to conceive of copyright as a battle between creators and consumers is misleading and damaging. People learn to create by seeing, imitating, experimenting, listening, practising, and watching; they learn from galleries or concerts, from the Internet, from family, and from school. When you think about it this way, you realize that creators are the most ardent consumers of the arts. They need ample and affordable access to the works of others, to libraries, and to education. They need, in copyright terms, users' rights.

The binary between creators and consumers is also problematic because it lumps together amateur creators with those who have had the talent and dedication to make cultural work into their main career. We think an appropriate term for professional creators is "cultural workers," because it recognizes the labour dimensions of their situation. Sometimes their interests dovetail with those of amateur creative types, and sometimes they don't. Meanwhile, consumers or users are both the market for works and a potential source of new works. The terms "consumer" and "user" have negative connotations that we think muddy the discussion. Perhaps the older term "audience" is more neutral. But the important point here is that receivers of works spend money, they learn, they seek out, they curate, and so even if they never put pen to paper or bow to string themselves, they need creators' rights, whether they know it or not. Thus while we speak of "creators' rights" and "users' rights," we do not map them onto discrete groups of people, the creators and the users. At different points in life, and in the context of different life circumstances or decisions, one set of interests or rights will loom larger than the other. But we all have a need for both.

Furthermore, the most powerful antagonist in copyright situations is often neither the creator nor the consumer but rather the corporation or the collective. Media corporations exploit cultural workers and profit from consumers. This isn't meant as an insult: it's just how business works. When corporations behave badly with regard to cultural workers, consumers pursuing other options say things like "Oh well, the money never would have gotten to the musician anyway," and cultural workers suffer double injuries. When collectives behave badly with regard to educational institutions, cultural workers often support the collectives, even if the benefit to them is not clear. This is all fairly dysfunctional, and it is important that wherever we stand we learn to be more clear about the structure of the cultural industries to avoid collateral damage to those who actually share some of our interests. We would add that we do not mean to tar all publishers, labels, or collectives with the same brush. To say "Those publishers, they're just out to rip us off" is to unfairly discount the value added by cultural mediators (be they editors, recording engineers, curators, designers, or rights clearance staff) and to unhelpfully lump together huge multinational profit centres and tiny local literary presses.

This is not an easy time for cultural industries. Creators and publishers find themselves squeezed and sometimes even pummelled by new media pathways, content, and tools and reduced government funding. In some professions, like journalism, it really does seem like the sky is falling. But we would argue that copyright is neither the main culprit nor a very effective solution. Media consolidation and new technologies are more direct causes of the challenges. We hope that attention will be turned to new ways to deliver licensed content that can coexist with and complement new modes of creation and free circulation. Some creators grieve income lost to unauthorized copying, but iTunes sales and Netflix subscriptions continue to grow in Canada, and many creators are finding the Internet indispensable to the making, marketing, and distribution of their work. We hope those trends will continue.

It is not an easy time for education and libraries either. Prices of educational materials climb and student numbers grow, while revenues, in many cases, are frozen or shrinking. But these institutions now have in hand a

New to Copyright in 2012

From Parliament:
- performers' moral rights
- photography treatment standardized with other works
- fair dealing purposes include satire, parody, education
- new and updated educational exceptions
- new consumer exceptions: time shifting, user-generated content, etc.
- circumvention of digital locks prohibited
- Internet service providers required to give users notice of alleged infringement
- reduction in statutory damages for non-commercial infringements

From the Supreme Court:
- affirmation of users' rights, ample fair dealing
- affirmation of technological neutrality principle
- clarification of scope of owners' rights

powerful affirmation of fair dealing from the Supreme Court. Statutory damages have been reduced. It is time for schools and libraries to make the most of the favourable situation. And that includes not only trying to save or make money, but also sharing their wealth of knowledge, expertise, and collections with a wider public. Digitization projects, for example, provide resources for students, community members, and creators alike, and can help to close the gap between those who identify primarily as creators and those who identify primarily as consumers. Educational institutions now have the opportunity to justify the faith put in them by Parliament and the Supreme Court.

• • •

This book has a strong Canadian focus because Canadians are short on practical and accurate information about what we can and can't do within the framework of our own copyright law. Canadians tend to know more about U.S. law. Copyright litigation in the United States is more frequent and often more notorious, U.S. law has moved fast and controversially in a maximalist direction, U.S. copyright warnings and ads preface almost every movie and DVD we watch, and U.S. public interest watchdogs such as the Electronic Frontier Foundation are fighting back with vigour. But there are many important differences between Canadian and U.S. copyright law. We need to know those differences. Canadian law is what we live under, whatever the origin of the materials in question.

We have organized the book into four parts. In Part I we survey the major philosophical and economic justifications for copyright (chapter 1) and Canadian copyright's origins in British, French, and U.S. traditions (chapter 2). While a discussion of philosophical concepts such as utilitarianism may seem intimidating, time spent here may help you to place and assess the copyright claims you hear around you on an everyday basis. The thumbnail early history of copyright has many fascinating resonances with present-day problems and controversies. Canadian copyright law particularly has always been caught between international forces, and it still is. It helps to know where we've come from.

Part II takes us to and through the Copyright Act, with focus on the amendments from 2012. Reading the Act systematically and understanding its context in case law provides the necessary groundwork for analyzing and crafting solutions in particular situations. In this spirit, we survey (chapter 3) the requirements for copyright to subsist in a work, or in some other subject matter, and look at the differences between different classes of works, explaining certain basic requirements such as originality and fixation in a tangible medium. We then enumerate (chapter 4) the rights held by an owner of copyright. While people usually think of copyright as the right to prohibit the making of copies, it is really much broader than that. Chapter 5, on Users' Rights, explains the scope and details of limitations on owners' rights. Copyright law has historically privileged owners' rights to the detriment of users' rights, but here we review a series of recent Canadian court cases that give much more weight to the rights of users of copyrighted materials. Chapter 6 describes bodies unique to the Canadian situation: copyright collectives and the Copyright Board. Chapter 7 addresses the question of who owns copyright—it isn't always the author. Finally, we look at what happens if you or someone else wants to act against infringement (chapter 8). This chapter covers practicalities such as the difference between civil and criminal infringement, cease and desist letters, small claims court, and statutory damages.

Part III covers more specific terrain, considering the issues that copyright presents for people creating and using particular media, or working in certain creative communities, institutions, or industries: from music and digital media through film and photography, visual arts, and craft and design, to the areas of journalism, education, and libraries and museums. In each case we identify special circumstances, real-life examples, and important case law, exploring sometimes thorny issues of both owners' rights and users' rights. You can dip into these chapters according to your particular needs and interests. They don't have to be read completely or in order, but they do presume that you've read Part II and are comfortable with the basic terms, principles, and building blocks of copyright.

Needless to say, we can't anticipate or answer all of your copyright questions. We don't specifically address the full range of artistic or craft practices—

dance, theatre, and video-game design, for example, are areas we've yet to delve into. And in a book of this nature we can't cover all the myriad details of the Copyright Act and case law. When it comes to a particular practice, the law evaluates each fact situation individually, and it isn't often possible to extrapolate with certainty from an analogous situation. So if you have a worrying legal dilemma, you will need to conduct further research or consult a lawyer. But if you have read this book first, you will at least be armed with basic terminology and good questions. You might even get some pleasure in seeing the look of surprise on the lawyer's face when you ask, "But what about section 29.24?"

Part IV outlines some alternatives and counterparts to copyright, from Indigenous customary law to citation economies, the open source movement, and public funding. We argue that copyright has too prominent a role as *the* solution in cultural policy when in fact it functions best as only one policy tool among many others. In the final chapter, we present a few areas to watch for future developments in Canadian copyright.

We hope that you will find some answers to your questions in this book. But even more, we hope that once you have read the book, you will be able to practise copyright attuned to the big issues of culture and democracy that it raises.

• • •

The authors of this book come to a common interest from different directions. One of us, Sam Trosow, is an Associate Professor at the University of Western Ontario; he is jointly appointed in the Faculty of Law and in the Faculty of Information and Media Studies (FIMS). He previously worked in California as a practising lawyer and later as a law librarian. His academic research focuses on information policy and political economy of information and knowledge: that is to say, where information and knowledge come from, how they circulate, and how they intersect with political and social processes. Beyond his work on copyright in the digital environment, he also has strong interests in legal theory and in the role of libraries as public information services.

The other author, Laura Murray, is an Associate Professor in English and Cultural Studies at Queen's University. Her background is in Aboriginal studies and eighteenth- and nineteenth-century American literature. She was first drawn to learn about copyright issues upon hearing Indigenous artists speak of the mismatch between copyright and their way of thinking about cultural custodianship. As she followed early twenty-first-century American debates over the constitutionality of copyright reform and the effects of digital technology on culture, she felt that a Canadian literary critic could offer something missing from the discussion. In 2003, concerned that there seemed to be few sources of information about emerging legislative reform in Canada, she started the website www.faircopyright.ca, which ran until 2010. Her work continues to focus on how ordinary people think about, or don't think about, copyright: with Tina Piper and Kirsty Robertson, she is coauthor of *Putting Intellectual Property in Its Place: Rights Discourses, Creative Labour, and the Everyday* (Oxford University Press, 2013), which develops many of the ideas first formulated in the "Copyright's Counterparts" chapter of this book.

The book emerges from an enormously rich and dynamic conversation among a wide range of people. For assistance with the second edition, we would like to thank Martha Rans for advising us, with great patience and persistence, on issues important to artists. She will no doubt take issue with some parts of this book, but her comments certainly made it better. Eli MacLaren and Myra Tawfik offered crucial corrections, updates, and enrichments to the history chapter. Jean Dryden and Mark Swartz provided much-appreciated advice on library and university contexts. Tina Piper, Kirsty Robertson, and Jane Anderson continued to expand Laura's horizons in thinking about law, art, and power. And an inspiring symposium convened by Ariel Katz just prior to the delivery of the manuscript made sure we were up to date on a range of implications of recent legal developments: thanks to all the participants.

Of course, the second edition builds on the first, so we reiterate our earlier thanks from the 2007 edition. On the law and policy end we thank Jody Ciufo, David Fewer, Michael Geist, Paul Jones, Elizabeth Judge, Howard Knopf, Wallace McLean, Russell McOrmond, Ira Nadel, Myra Tawfik, Paul Whitney, and all the members of the faircopy listserv. The thoughts and expertise of artists and writers were important to the genesis of the book: we thank especially Karl Beveridge, Susan Crean, John Degen, Richard Fung, John Greyson, Christopher Moore, and the participants of Copycamp, September 2006, even and especially when our opinions differed from theirs. Comments and inquiries from many Canadians that came in via the faircopyright website were invaluable prompts about what ordinary people wanted to know about copyright. Special thanks to the artists and others who agreed to be interviewed on their experiences with copyright, and to Kirsty Robertson, Linda Quirk, and Shannon Smith for research and editorial assistance. Jane Burkowski proved a tolerant as well as talented illustrator, and she has updated and invented illustrations anew for the second edition, even as she is defending her PhD thesis at Oxford.

Laura's grants from the Social Science and Humanities Research Council and Sam's from the Graphics, Animation and New Media (GRAND) NCE provided much-appreciated research funding; Laura is also grateful to Queen's University for a sabbatical that enabled the writing of the first edition. Working with Paul Eprile, Jennifer Tiberio, and Robert Clarke of Between the Lines was a pleasure in 2007; we now add thanks to Amanda Crocker, Renée Knapp, Matthew Adams, and Paula Brill at BTL and to our very able editor Tilman Lewis, our designer Gordon Robertson, and our indexer Martin Boyne. Our families, friends, and communities continue to keep us thinking, eating, playing, and doing. Peter Murray especially has been a patient and supportive interlocutor. And last but not least, we thank Dropbox: we couldn't have written the book without it!

Laura Murray and Samuel Trosow

PART I

IDEAS

1. COPYRIGHT'S RATIONALES

Copyright is so entrenched in popular thinking about the production and dissemination of culture that we may think of it as natural or inevitable. We may even drape it with mystical ideas about the creative process. To be sure, authors and artists have always had a special connection to their work. The seventeenth-century poet John Milton wrote that books "preserve as in a vial the purest efficacy and extraction of that living intellect that bred them." An anonymous author declared to the British Parliament in 1735, "If there be such a Thing as Property upon Earth, an Author has it in his Work."[1]

These claims were made, however, as polemical assertions in the midst of raucous debate, not as statements of established fact. In exalting authors as sources or owners, Milton and the anonymous author spoke against a common sense of their time, according to which artists were honoured as custodians and animators of collective tradition. In ancient and traditional cultures worldwide, from Greece to New Orleans to Haida Gwaii, the artist does not create but re-creates, does not own but feeds. Artistic and intellectual production understood in this collective way tends to be supported by patronage rather than by a system of individual rights or property. It is important, therefore, to clarify copyright's specific logic.

Established Philosophies of Copyright

Why should copyright holders have exclusive rights in their works? Copyright laws rest on two major lines of philosophical justification: *rights-based theories* and *utilitarianism*. Both of these approaches have advantages and limitations, and both of them are explicitly or implicitly represented in today's copyright debates.[2] The economic analysis that holds sway in many quarters today can be seen as a descendant of both lines of thought.

Rights-Based Theories

Rights-based theories are rooted in ideas of natural law. Proponents of natural law believe that the law exists independently, separate and apart from legislation that has been posited by any particular state. While natural law may be associated with a religious world view, it can also appeal to an abstract moral authority, such as justice. The principles of natural law are expressed in documents such as the Magna Carta and the French Declaration of the Rights of Man. The claim from the American Declaration of Independence "that all men are created equal, that they are endowed by their Creator with certain unalienable Rights" is a good example of natural law philosophy. More generally, the idea of human rights is derived from a natural law approach: rights come from "nature or nature's God," as the Declaration of Independence puts it, not from a particular ruler or government.

A natural law approach to property would hold that each person has a natural entitlement to their person and to the fruits of their labours. The most well-known expositor of this philosophy is John Locke, who in his *Second Treatise of Government* (1690) set out a theory that justifies the private appropriation of public resources. While Locke was writing about the appropriation of physical resources (that is, land and things), his work has come to be applied to intellectual labour as well. Locke begins with the premise that "the *Labour* of [a person's] Body, and the *Work* of his Hands, we may say, are properly his." Then he says that whatever a person "removes out of the State that Nature hath provided, and left it in, he hath mixed his *Labour* with, and joyned to it something that is his own, and thereby makes it his *Property*."[3]

In a Lockean view of copyright, the labour supplied by the author provides a justification for a claim to exclude others—even if the author is working with materials previously available to all. A claim that copyright ought to be perpetual could also be justified by reference to Locke, because property rights in physical resources are perpetual.

But Locke also specified two limitations on the right to appropriate from the commons. First, he stated that the appropriation must leave as much and as good for others; second, he did not consider ownership legitimate when individuals appropriated more than they could use.[4] Locke was also explicitly opposed to perpetual copyright.[5] Thus, whether we are talking about tangible property or intellectual property, Locke may provide justification both for owners' rights and for limitations to them.

On some level, many people may think of copyright as a natural right because it just seems fair that authors should hold rights in work they have created. But the courts, in the Anglo-American tradition, do not see it this way. In the seventeenth century, English courts held that Acts of Parliament were subject to the constraints of natural law, often understood to be embodied in common law, or the accumulated collection of precedent from specific legal cases. In *Dr. Bonham's Case* (1610), the court said that the "common law will control Acts of Parliament, and sometimes judge them to be utterly void: for when an Act of Parliament is against common right and reason, or repugnant, or impossible to be performed, the common law will control it, and adjudge such Act to be void." But after 1688, Acts of Parliament were thought to be supreme: in other words, the law was understood to lie in what the government had expressly promulgated, enacted, or posited. In the

Are movie studios' copyrights in movies based on fairy tales legitimate, according to Lockean thinking?

Yes: the studios created private property by taking a story from the public domain and adding their labour to that story. But Locke was talking about a world of limitless resources. In his *Second Treatise of Government* (chapter 5, section 33), he wrote: "No Body could think himself injur'd by the drinking of another Man, though he took a good Draught, who had a whole River of the same Water left him to quench his thirst." So this raises the question of whether fairy tales are a limitless resource. Does a studio's taking of them leave less for others? Copyright law has developed distinctions between rival and non-rival goods, and between ideas and expression, in order to answer questions such as these. While a studio can own the rights to its version of the story, copyright law holds that such ownership only extends to the new elements that the studio adds. The story itself has to be left free for others to use as well. This is why the two 2012 Snow White movies don't raise any legal issues, even given the existence of the sublime *Snow White and the Seven Dwarfs* from 1937; in Lockean terms they are merely "draughts from the same water."

realm of copyright, this "positive law" viewpoint was confirmed in the 1774 case *Donaldson v. Becket*.[6] In this case, a divided House of Lords affirmed the limited copyright term of the Statute of Anne over claims of common-law perpetual copyright, rejecting the notion of a "natural" copyright separate and apart from the statute.[7] Thus, while today's justifications for copyright law are often rooted in the thinking of natural law, Anglo-American law now operates predominantly according to positive law principles.

Still, rights-based or natural law theories do continue to play a more central role in the civil law systems that originated in Continental Europe, brought to Canada through French law.[8] Civil law systems place more emphasis on the individual rights of the author as a person, and tend to view

> [Philosopher Robert] Nozick asks: If I pour my can of tomato juice into the ocean, do I own the ocean? Analogous questions abound in the field of intellectual property. If I invent a drug that prevents impotence, do I deserve to collect for twenty years the extraordinary amount of money that men throughout the world would pay for access to the drug? If I write a novel about a war between two space empires, may I legitimately demand compensation from people who wish to prepare motion-picture adaptations, write sequels, manufacture dolls based on my characters, or produce t-shirts emblazoned with bits of my dialogue? How far, in short, do my rights go?
>
> —William Fisher, "Theories of Intellectual Property," 188–89.

copyright as an extension of the personality of the author. Canadian law represents a blending of English and French traditions, and Supreme Court cases in particular often reflect a combination of the two.

Utilitarianism

Utilitarianism is another major stream of justification of copyright. As a broad school of thought, utilitarianism is generally attributed to the nineteenth-century English philosopher Jeremy Bentham. According to Bentham, people can make decisions in a situation of competing interests by measuring the total amount of happiness produced. "A measure of government," he wrote, "may be said to be conformable to or dictated by the principle of utility, when...the tendency which it has to augment the happiness of the community is greater than any which it has to diminish it."[9] The so-called copyright clause of the U.S. Constitution might be taken as an example of utilitarianism: it does not appeal to a higher power, as in natural law thinking, but rather empowers Congress to enact intellectual property laws as a tool for general benefit—that is, "to promote the progress of science and the useful arts." While Canada's copyright principles are not articulated at the constitutional

> Like many since, the eighteenth-century English writer Samuel Johnson combined natural law and utilitarian thinking in his approach to copyright:
>
>> There seems . . . to be in authours a stronger right of property than that by occupancy; a metaphysical right, a right, as it were, of creation, which should from its nature be perpetual; but the consent of nations is against it, and indeed reason and the interests of learning are against it; for were it to be perpetual, no book, however useful, could be universally diffused amongst mankind, should the proprietor take it into his head to restrain its circulation. . . . For the general good of the world, therefore, whatever valuable work has once been created by an authour, and issued out by him, should be understood as no longer in his power, but as belonging to the publick.
>
> Source: Johnson quoted in Boswell, *Boswell's Life of Johnson*, 546.

level, our courts and legislators have often and increasingly used a rhetoric of public or national interest that could be said to be utilitarian.[10]

Economic Analysis

In today's debates, copyright is most often justified in economic terms: we are living in a knowledge-based economy, the claim goes, and we need a particular vision of copyright to drive that economy. Classic economic analysis of copyright law rests on three general assumptions: that the free market system is the appropriate allocation device to guide the creation and dissemination of "intellectual and information goods"; that these goods will be underproduced without a guarantee of sufficient market-based financial incentives to creators and owners; and that the expansion of exclusive intellectual property rights is necessary to protect these market-based incentives from being undermined by acts of appropriation.

Within the limitations of these assumptions, economic analysis seeks to promote the efficient allocation of resources in a market setting. In its sacralization of property rights, it is underpinned by natural law philosophies, but it is also essentially utilitarian in nature, in that it recognizes the existence of a trade-off between limiting access to works and providing economic incentives to create works. After all, an economy in which every single transaction with a copyright work was monetized or metered in some way would carry great financial and bureaucratic costs, which might slow down its growth (economists call these "transaction costs"). The trade-off is often referred to as the balancing of interests between the rights of owners and the rights of users.

The Internet does lower the cost of copying and, thus, the cost of illicit copying. Of course, it also lowers the costs of production, distribution, and advertising, and dramatically increases the size of the potential market. Is the net result, then, a loss to rights-holders such that we need to increase protection to maintain a constant level of incentives? A large, leaky market may actually provide more revenue than a small one over which one's control is much stronger. What's more, the same technologies that allow for cheap copying also allow for swift and encyclopedic search engines—the best devices ever invented for detecting illicit copying. It would be impossible to say, on the basis of the evidence we have, that owners of protected content are better or worse off as a result of the Internet. Thus, the idea that we must inevitably strengthen rights as copying costs decline doesn't hold water. And given the known static and dynamic costs of monopolies, and the constitutional injunction to encourage the progress of science and the useful arts, the burden of proof should be on those requesting new rights to prove their necessity.

—James Boyle, "The Second Enclosure Movement and the Construction of the Public Domain."

Such cost-benefit analysis is open to criticism on a number of grounds. The losses that come from limiting access are not as susceptible to precise measurement as are the financial benefits accruing to the owners of exclusive copyright interests. And the degree to which financial incentives drive creativity is also not easily measurable in many areas of endeavour. Thus the behaviour of both creators and consumers may be less well addressed by economic analysis than the behaviour of vendors and producers.[11] The balancing approach also tends to divide the world into owners and users, when most of us are both. And it does not seem to adequately consider how different stakeholders come to the table with different resources, different values, different backgrounds, and different levels of political power. But while it may be argued that the discourse of "balancing of interests" fails to address several problems, it does usefully frame both owners' and users' rights in pragmatic terms as parts of a dynamic creative economy rather than as matters of fairness that are easily treated with lip service and then ignored.

This approach is also better than a one-dimensional argument that protections are good and more protections are better. The claim is often made that copyright protections need to be expanded because of changes in technology, or because new cultural practices threaten existing business models. But looking at copyright only from the standpoint of protections overlooks the reality that one person's additional rights are just further restrictions for someone else. Rather than thinking about rights in a vacuum, we prefer to think also about the corresponding duties and disabilities that the rights impose on others. In other words, it makes just as much sense to speak of "copyright restrictions" as of "copyright protections."

Intellectual Creations as Public Goods

So far we have introduced two major paths of philosophical justification for copyright, and suggested how they underlie modern economic analysis. While we pointed out some of their pitfalls, we generally followed the tendencies of both approaches to gloss over the distinction between tangible goods (land, chattels, widgets) and intangible goods (expression, knowledge, information). However, the differences between tangible and intangible goods are fundamental, and any fully convincing justification of copyright (or, for that matter, patent, although we will not get into that here) must recognize these differences. Talk about the importance of flows of information and knowledge is ubiquitous: Canadians are constantly being told that we live in an information society, and that we must take the lead in innovation. But little attention has been paid in the policy context to understanding the nature and characteristics of information, ideas, knowledge, and human expression.

Here's an example of the issues at play. A book is *personal property*. It's tangible, which is to say, translating the Latin root *tangere*, touchable. You can hold it in your hand and if someone takes it, you no longer have it. The words, illustrations, and design in the book are *intellectual property*. They are intangible, in the sense that they were probably generated on a computer and could be embodied as computer code or as an audio file and still be themselves. And then there's an even more intangible layer of book contents: the ideas or facts within it, which, as we shall see in chapter 3, cannot be owned.[12]

As we have seen, economists speak of intangibles, both ideas and expressions of them, as "intellectual and information goods," which they categorize as public goods as opposed to private or tangible goods. Public goods exhibit two major differences from private goods: they are generally *non-rival in consumption*, and they do not inherently possess *exclusion mechanisms*. It's worth examining these two concepts in some detail.

If a good is "rival in its consumption," it is depleted or used up when one person consumes it. Physical consumer goods that populate store shelves are rival in consumption. When a widget is purchased it is no longer on the shelf for the next shopper. Depletable energy resources are another classic example of rivalry in consumption. When we say that public goods are non-rival in consumption, we mean that the consumption of the good by one person does not reduce the amount of the good available for consumption by others. If you walk down a street illuminated by a street light, the light is not depleted because you enjoyed its benefit. The bulb in the lamp

> If nature has made any one thing less susceptible than all others of exclusive property, it is the action of the thinking power called an idea, which an individual may exclusively possess as long as he keeps it to himself; but the moment it is divulged, it forces itself into the possession of everyone, and the receiver cannot dispose himself of it. Its peculiar character, too, is that no one possesses the less, because every other possesses the whole of it. He who receives an idea from me, receives instruction himself without lessening mine; as he who lights his taper at mine, receives it without darkening me.
>
> — Thomas Jefferson to Isaac McPherson, Monticello, 13 August 1813, in Jefferson, *The Writings of Thomas Jefferson*, 13, 333–34.

will be depleted through use and is itself a private good with rivalry in consumption. But the service of street lighting is a public good and exhibits non-rivalry in consumption. The act of breathing does not significantly reduce the air available for everyone else, so it, too, is non-rival in consumption. (Locke said the same about water, so we can see that goods can change, depending on circumstance, from non-rival to rival or vice versa.)

In the context of copyright analysis, we can distinguish a book or a DVD (physical goods with rivalry in consumption) from the information and expression contained in the book or DVD. Until recently, information and expression were necessarily distributed in physical containers, so the differences between rivalry and non-rivalry in consumption were not as noticeable as they are today. But with advances in digital technology, content is now routinely severed from its container. A digital file is non-rival in consumption and can be distributed to ten thousand persons just as well as ten. One could even say that the essence of information as information—like language as language, or images as images—is that it is non-rival in consumption.

There are certainly exceptions to this general observation. For example, hot market information and other types of proprietary data might become less valuable with wider distribution.[13] But we could also note that much information or expression becomes *more* valuable as more people use it, by the phenomenon known by economists as "network effects." Facebook provides an apt analogy: it would not be very valuable to you if you were the only person on it, but the more people using it, the more value it has. A similar thing happens in the cultural marketplace with bestsellers, fads, and trends.

The second aspect of a public good that distinguishes it from a private good is that it does not have an "exclusion mechanism." A tollgate is an example of an exclusion mechanism. So is a cash register: when you go to the store, you don't get to enjoy a new shirt or bicycle unless you pay for it. Public goods are different. Anyone can use them, regardless of whether they express a preference for them in the marketplace. People who walk down a street at night get the benefit of the street light whether or not they helped pay for it. No shield emerges to block the light from those who have not paid taxes in that jurisdiction (or at least not yet: maybe someone will think of a way to do it). National defence, policing, roads, and schooling are other common examples of goods that lack an exclusion mechanism. You enjoy the benefits of national defence expenditures whatever your opinion on how tax revenues should be spent.

Whether or not a good has an effective exclusion mechanism can be a question of public policy, a question of technology, or both. The law of theft

Table 1. Comparison of Pure Public Goods and Pure Private Goods

	Consumption	Exclusion Mechanism
pure public good	non-rival (joint) consumption; use does not result in depletion of the good	exclusion mechanism is not present
pure private good	rival consumption; use results in depletion of the good	exclusion mechanism is present

is an exclusion mechanism that has long been imposed as a matter of public policy. It is against the law to take an item out of a store without paying for it, and it is against the law to sneak into a theatre without buying a ticket. The exclusion mechanism may also be a technological device. The automated tollgate is an older such technology, and consumers are now becoming familiar with a vast array of new digital locks or gateways, known as "technological protection measures (TPMs)." But exclusion mechanisms are often hybrid; that is, the law often acts to reinforce a technological exclusion mechanism. Think of cable television. It used to be that television airwaves were pure public goods. By turning on your television and viewing a broadcast, you were not depleting the airwaves available for others to enjoy. Cable companies introduced an exclusion mechanism: you had to pay to get the system hooked up. If you fix the cable box so that you can view programs without subscribing to the service (or create a device to do so), you are likely to be in violation of a law and subject to sanctions. The same double exclusion mechanism has, with recent changes to Canadian copyright law, been layered onto TPMs (see chapter 10).

We have seen how intellectual goods are inherently non-rival in consumption, as they are not naturally subject to an exclusion mechanism. The container holding the information (the book, the DVD) is subject to an exclusion mechanism and rivalry in consumption, but the information contained therein is not. Public goods present a problem for market-oriented economists because, if an item has public good characteristics, people will be able to use and enjoy it without having to pay for it. Lack of exclusion means you can obtain the benefit of the good whether or not you are willing to pay for it. The price system, which is based on rules of supply and demand, cannot operate for public goods, and we have in this an instance of what economists call "total market failure."

While many people see the public goods quality of digital information and expression as an exciting phenomenon, mainstream economists and large-content owners see public goods as a problem that needs to be solved. They desperately need the price system to work. A fix is needed, and the cure is to create some sort of exclusion mechanism. In the case of intellectual *because info is a non-rival good*

goods, the laws of intellectual property can be layered on top of technology to create scarcity and impose constraints on free flows of information.

Creators may also seek such fixes, as they face new challenges to controlling the circulation of the fruits of their labour and talents. Just because their work has now become non-rival doesn't necessarily mean they wish to give it away. They may figure they deserve to be paid for their labour like everyone else, and they may find digital locks attractive. Digital locks underlie paid music download services such as iTunes, and one could pursue similar business models for works such as poems or images: even in a context where free copies are available, one can find a market for convenient high-quality versions for a price. One problem, however, is that exclusion mechanisms tend to work insofar as the market works. But the market doesn't work terribly well for many creators—just as it often hasn't for other important resources like water and electricity. It has often been effective to pay for the arts as we pay for street lights, that is to say, through taxes. If limiting access to creative products is either undesirable or impossible, or if it doesn't support the range of work we may wish to foster, maybe it's appropriate to treat them and pay for them more like other public resources, at least in part.

The overall point here is that artificially created exclusion mechanisms are powerful policy tools. They may well be justifiable on natural law or utilitarian grounds: we may say, for example, that it isn't fair that authors not be paid for their work, or that it is in the public interest that they be paid. Or we may think that only the individuals who need a certain good ought to pay for it. But exclusion mechanisms should be used carefully, because they may not always foster creativity or innovation, and they have the potential to unduly restrict the transfer of information and knowledge.

2. COPYRIGHT'S HISTORIES

Where does Canadian copyright law come from? Like so many Canadian institutions, it is the product of a long history of imposed and adapted British law and competing French traditions, complicated by the weighty proximity of U.S. markets and cultural influence. British, French, and U.S. law are all quite different, and to understand Canada's situation fully we need to know something about the early histories of each of them, as well as the history of copyright in Canada through the twentieth century. Within copyright's early history—and in particular within the eighteenth- and nineteenth-century book trade—we find familiar versions of many of today's debates and dilemmas, including the challenges posed by digital technologies. While we may or may not resolve these issues in the same way now, the historical comparisons can alert us to a range of possibilities and forces that we may not see by gazing only at our present situation—the international dimensions of which are laid out at the conclusion of the chapter.

First, though, there are three overall observations to keep in mind. One is that copyright's history is quite short—action-packed, but short. Art created before the late eighteenth century was financially enabled not by the concept of owners' rights, but by some version of patronage, commission, or employment. Second, although copyright does indeed benefit many authors, authors have not been its main concern or driving force. Throughout history, it is the larger book trade—today the cultural industries or tech sector—that has demanded expansion of copyright, often using the rhetoric of authors' rights to do so. Third, copyright is a pragmatic policy tool that exists in widely differing forms. How sharp or blunt the tool should be, how broad its application should be, whom it should benefit, and how long it should exist— these have been and will continue to be matters for public policy to sort out.

Canada's Three Copyright Legacies

The United Kingdom

Modern Anglo-American copyright is usually said to begin with Britain's 1710 Statute of Anne, titled in full "An Act for the Encouragement of Learning, by Vesting the Copies of Printed Books in the Authors or Purchasers of such Copies, during the Times Therein Mentioned."[1] The Statute of Anne marked a departure from the printing regulations of the 150 years preceding it. With the introduction of the printing press to England in the fifteenth century, the Crown had feared the spread of seditious works. To aid in controlling a dangerous new technology, it granted a publishing monopoly in 1557 to the Stationers' Company, a group of London printers and booksellers who could be relied upon to censor works in exchange for large profits. During this period, authors' main support came through patronage: their expenses would be lightened by wealthy supporters. They could sell their "copy" once, but could not benefit from or set the terms for its reprinting.[2]

During and following the Civil War of the mid-seventeenth century, the licensing system fell into disarray, and in 1695 Parliament let the Licensing

Act lapse. This was no act of political openness; as Joseph Loewenstein puts it, "The licenser's judgment was . . . to be displaced by the more methodical constraints of the laws of libel, seditious libel, and treason."[3] It is partly out of this situation that a sense of authors' rights began to emerge. Novelist (and journalist) Daniel Defoe declared, "If an Author has not a right of a Book, after he has made it, and the benefit be not his own, and the Law will not protect him in that Benefit, 'twould be very hard the Law should pretend to punish him for it."[4]

During the unregulated period, printing enterprises began to spring up in the provinces, to meet growing market demand. Panicked at the loss of its monopoly, the Stationers' Company wrote to Parliament with the dire warning that, as copyright historian Mark Rose frames it, "If Parliament failed to confirm [their] literary property, thousands of mechanics and shopkeepers would be deprived of their livelihoods, and 'Widows and Children

In 1663 Sir Roger L'Estrange helpfully laid out the range of parties involved in the production of a book:

> The Instruments for setting the work [of promotion] afoot are These. The Adviser, Author, Compiler, Writer, Correcter, and the Persons for whom, and by whom; that is [to] say, the Stationer (commonly), and the Printer. To which may be Added, the Letter-Founders, and the Smiths, and Joyners, that work upon Presses. The usual Agents for Publishing, are the Printers themselves, Stitchers, Binders, Stationers, Hawkers, Mercury-women, Pedlers, Ballad-singers, Posts, Carryers, Hackney-Coachmen, Boat-men, and Mariners.

In today's publishing, music, film, and broadcast industries, just as many professions have a stake in the business, and hence in copyright.

Source: McKeon, *The Secret History of Domesticity*, 51.

who at present Subsist wholly by the Maintenance of this Property' would be reduced to extreme poverty."[5]

In 1710 Parliament did move to regulate the book trade, but without reinstating the Stationers' monopoly and perpetual rights. In the Statute of Anne it limited the term of copyright to fourteen years, renewable for another fourteen if the author was still living. It allowed parties outside the Stationers' Company—authors and their assignees—to own those rights. While modern copyright Acts have added complexity and nuance to copyright's operation, applied it to media other than books, and extended its term (tying it to the length of the author's life), the core of copyright comes down to us from the Statute of Anne. That is, the author is given a monopoly to exploit the work, and to restrict others from doing so, for a limited time.

The limited time period in the Statute of Anne presented difficulties for booksellers used to a perpetual monopoly.[6] They accepted the idea of authors' rights fairly quickly, partly because the courts made it clear that publishers' rights were based on authors' rights, and partly because rhetorically authors' rights were (and remain) a more powerful rallying cry than publishers' or booksellers' rights.[7] Publishers even argued that authors' rights were perpetual under common law, and it was not until 1774 that this line of argument was rejected. In *Donaldson v. Becket*, the House of Lords ruled that the Statute of Anne cancelled any existing common-law copyright.[8] This decision leaves us with the principle that has underpinned Anglo-American copyright ever since: that copyright is a creature of statute alone, not common law.

The United States

As U.S. legislators established a body of law for their new nation, they started with the British Statute of Anne as a model, but soon developed different legal principles better suited to their particular stage of cultural and economic development. Although Canada's law is based on British law, we need to understand early U.S. law to grasp the origins of Canadian copyright in the nineteenth century; developments in Canada in this period lay

> . . . for the encouragement of learned men to compose and write useful books; may it please your Majesty, that it may be enacted . . . That from and after [10 April 1710], the author of any book or books already printed, who hath not transferred to any other the copy or copies of such book or books . . . shall have the sole right and liberty of printing such book and books for the term of one and twenty years, to commence from . . . [10 April 1710], and no longer; and That the author of any book or books already composed, and not printed and published, or that shall hereafter be composed, and his assignee, or assigns, shall have the sole liberty of printing and reprinting such book and books for the term of fourteen years. . . . Provided always, That after the expiration of the said term of fourteen years, the sole right of printing or disposing of copies shall return to the authors thereof, if they are then living, for another term of fourteen years.
>
> — Statute of Anne, 1710.

very much in the shadow of the burgeoning U.S. book industry and its particular legal and philosophical underpinnings. Early U.S. copyright is also interesting because of the contrast it presents to present-day U.S. law.

Certain key elements of an approach to copyright are embedded in the U.S. Constitution. Article I, section 8, grants Congress power to enact legislation "to promote the progress of science and useful arts, by securing for limited times to authors and inventors the exclusive right to their respective writing and discoveries." This clause enables various forms of intellectual property law, including patent and copyright.[9] Note the similarity between the language of "promotion" and the Statute of Anne's stated purpose of "Encouragement": both are examples of the utilitarian justification for copyright. Exclusive rights are not in themselves the constitutional goal: they exist to enhance a long-term goal of public benefit, that is, the promotion of science and the arts.[10] Beyond the copyright clause of the Constitution, another section speaks to related issues. The framers were concerned that government not be permitted to undermine the robust exchange of ideas necessary for a democracy. In response to the history of censorship in England, they developed the First Amendment to the Constitution, which guarantees freedom of speech.[11]

The first U.S. Copyright Act, set down in 1790, is notable for its refusal to grant copyright protection to non-American works or authors.[12] As Meredith McGill argues, this omission was not an oversight but a matter of principle, and Congress repeatedly refused to grant copyrights to foreign authors through the century that followed. McGill writes, "Not only was the mass-market for literature in America built and sustained by the publication of cheap reprints of foreign books and periodicals, the primary vehicles for the circulation of literature were uncopyrighted newspapers and magazines."[13] Publishers at the time lamented the disadvantage that this regime imposed upon U.S. authors, who had to compete with cheap British blockbusters for audiences.[14] British authors, notably Charles Dickens, were in turn incensed at the cheap circulation of their work in the United States. However, the availability of cheap books clearly contributed to the building of a hungry and educated American reading audience.

As U.S. markets developed, the rhetoric of authors' rights did begin to emerge in the country. But—partly because of the treaty's strong authors' rights provisions—the United States did not sign the 1886 Berne Convention (indeed, it would sign this treaty only in 1988), and it was not until 1891 that the U.S. government developed a reciprocal copyright arrangement with Great Britain. By 1903, as the United States came to be a major player in the international cultural trade, *Bleistein v. Donaldson* articulated an idea of natural authors' rights. In this case concerning copied circus posters, Justice Oliver Wendell Holmes of the U.S. Supreme Court argued that any image was "the personal reaction of an individual upon nature. Personality always contains something unique. It expresses its singularity even in handwriting, and a very modest grade of art has in it something irreducible, which is one man's alone."[15] As Carla Hesse points out, "Through the Holmes decision the rhetoric of authorial originality and natural rights . . . made its way into American jurisprudence at the very moment when America began to supplant Europe as the hegemonic global economic power."[16]

But even with this shift in perspective, U.S. lawmakers continued (with some limited exceptions) to steadfastly refuse any idea of rights vested by nature in the author.

France

Today it is a truism to say that Canadian copyright law derives from both French and British law—by which people usually mean that there is a (hopefully productive) tension between a "*droit d'auteur*" tradition deriving from natural law, and a "copyright" tradition that is more utilitarian in its philosophical underpinnings. However, utilitarianism has almost as long a history in France as it does in England. In a lively debate in the 1760s and 1770s, two philosophers—Denis Diderot and the Marquis de Condorcet—articulated contrasting views. Diderot held that products of the mind were even more like property than land itself, whereas Condorcet argued that literary property was "not a property derived from the natural order . . . it is a property founded in society itself. It is not a true right; it is a privilege."[17] The Crown agreed with Condorcet, but the Revolution revoked all existing legislation and protocol for the book trade. In 1791 Condorcet was involved in drafting a law that recognized works as property, but property that could only be held ten years past the death of the author. In 1793 the National Convention passed a version of this Act—based, like British law, on the idea of a limited property right. This statute governed copyright in France until 1957.[18]

In her survey of French case law and scholarship, Gillian Davies shows that throughout the nineteenth century authors' rights were not thought to arise directly out of natural law. The change in thinking seems to have occurred in the first half of the twentieth century. In the consulting and documentation associated with the new law of 1957, which was intended to codify existing practice, she finds a strong leaning towards natural rights thinking and very few mentions of the public interest. The rapporteur of the committee drafting the new law mentioned the public interest only to subsume it in authors' rights when he wrote that the goal of the new law was to "effect the synthesis of the author's rights and the interests of the public,

The most sacred, most legitimate, most unassailable, and if I may put it this way, the most personal of all properties, is a work which is the fruit of the imagination of a writer; however, it is a property of a kind quite different from other properties. When an author has delivered his work to the public, when the work is in the hands of the public at large, so that all educated men may come to know it, assimilate the beauties contained therein and commit to memory the most pleasing passages, it seems that from that moment on the writer has associated the public with his property, or rather has transmitted it to the public outright; however, during the lifetime of the author and for a few years after his death nobody may dispose of the product of his genius without consent. But also, after that fixed period, the property of the public begins, and everybody should be able to print and publish the works which have helped to enlighten the human spirit.

— Isaac Le Chapelier, Report to the French Parliament, 1791,
in Davies, *Copyright and the Public Interest*, 137.

in the preeminence of the creator."[19] Indeed, the very first article of the 1957 Act, still in place, explicitly repudiates a utilitarian philosophy of copyright:

> The author of a work of the mind shall enjoy in that work, by the mere fact of its creation an exclusive incorporeal property right which shall be enforceable against all persons. The legislator does not intervene to attribute to the writer, the artist, the composer, an arbitrary monopoly, under the influence of considerations of expediency, in order to stimulate the activity of men of letters and artists in the interest of the collectivity; the author's rights exist independently of his [the legislator's] intervention.[20]

French law does contain a number of exceptions that indicate a recognition of citizens' needs—it permits quotations for critical, informational,

polemical, scientific, or educational purposes; parody, pastiche, and cari-
cature; and some recordings of broadcasts for the purposes of preservation.
Its philosophical orientation, like its name, *droit d'auteur*, is clearly centred
on the author. But this single focus represents the ascendance of one of two
strands of French copyright thought—and when we speak of the French
tradition in Canadian law, we may be speaking of a tradition that developed
alongside Canadian law, rather than prior to it.[21]

The Beginnings of Copyright in Canada

Although the British North America Act of 1867 named copyright as an area
of Canadian federal jurisdiction, the British Parliament's so-called Imperial
Copyright Act of 1842 remained in force in Canada until a Copyright Act
passed in Ottawa in 1921 came into effect in 1924. This remarkably long
delay requires some explanation.

Copyright history in Canada before 1924 can be understood as a story
about the grip of British law and the weight of U.S. market forces on a cluster
of small colonies. As such, it offers a resonant foundation for thinking about
Canada's copyright interests today, during a time when we still have to craft a
position in the context of huge American cultural imports—although the
British legal framework has now been replaced with pressures from the
World Trade Organization (WTO) and the World Intellectual Property
Organization (WIPO).[22] The question remains now, as it was then: Are
Canada's interests the same as those of the major cultural exporters? If
not, to what extent is Canada free to develop its own policy directions and
mechanisms?

In the nineteenth century Canadians mainly read British books in U.S.
reprint editions. Books printed in Britain were expensive, with high shipping
charges as well, whereas U.S. printers were producing large numbers of
cheap, unauthorized reprints—which were not illegal under U.S. law. British
publishers were not amused at what they perceived as a lost British North
American market. Some in Britain were also concerned that Canadians'
access to U.S. books might be "sapping the principles and loyalty of the

Subjects of the Queen."[23] As a result, in 1842 the Imperial Copyright Act outlawed importation of reprints into Britain and its possessions, and the U.K. put a 35 per cent duty on U.S.-originated publications, provoking a huge outcry in British North America. After all, was it not Britain's responsibility to facilitate the education of its subjects? As author Susanna Moodie stated, "Incalculable are the benefits that Canada derives from cheap [U.S.] reprints of all the European standard works, which in good paper and in handsome bindings, can be bought at a quarter the price of the English editions."[24] In 1847 the Foreign Reprints Act permitted imports once again, for a duty of 12.5 per cent that was in practice seldom collected.

By the 1860s Canadian printers had joined in a discussion previously dominated by booksellers. Printing was flourishing because of demand for local newspapers, which Americans could not supply. With Confederation and broader economic development, printers saw the prospect of national markets, and they lobbied for a licensing scheme that would put them on a par with the Americans by allowing them to reprint British books without permission for a standard royalty. In 1872 the Canadian Parliament passed a Copyright Act containing such a provision.[25] But Canadian legislation required British approval, which was not forthcoming. As the Canadian printer John Lovell recalled after participating in a diplomatic mission to England, "The English publishers would not yield an inch. They said they would not allow any colonial to publish one of their books. Their ignorance of Canada was profound. They treated Canada as if it was part and parcel with the United States."[26] Canada did pass a Copyright Act in 1875, but it did nothing to allow unauthorized Canadian reprinting of British copyright works; rather, it gave British publishers a way to prohibit importation of U.S. reprints into Canada. Under the 1875 Act, Canadian publishing languished.[27]

As the Europeans moved towards the agreement that in 1886 became the Berne Convention for the Protection of Literary and Artistic Works, and the British and Americans laboured to come to a reciprocal agreement finally realized in 1891, Canadians grew more concerned. Although Berne did for the first time give authors publishing first in Canada copyright throughout the Empire and beyond, it threatened Canadians' access to reprints and

In 1843 a committee was struck by the Legislative Assembly of the Province of Canada to study copyright's effects on the Canadian book trade. It made it clear that what we now call "access" to printed material was necessary to the cultural and economic development of Canada, and it even argued that U.S. reprints were necessary to ensure Canadians' loyalty to the Queen:

2nd. that the free admission into this Province of American Reprints of English Works of Art and Literature, could not lessen the profits of English Authors and Publishers; because, although the reading population of the Province is great in number, yet the circumstances of the population generally are so limited in their means, that they are unable to enjoy English Literature at English prices; that owing to that inability to pay for such Work of Art and Literature there has never been a demand for those Works, and consequently no supply.

3rd. That the exclusion of American Reprints of English Literature, if possible, would have a most pernicious tendency on the minds of the rising generation, in morals, politics, and religion; that American Reprints of English Works are openly sold, and are on the tables or in the houses of persons of all classes in the Province; that a law so repugnant to public opinion cannot and will not be enforced; that were that exclusion possible, the Colonists would be confined to American literary, religious, and political Works, the effect of which could not be expected to strengthen their attachment to British Institutions, but, on the contrary, is well calculated to warp the minds of the rising generation to a decided preference for the Institutions of the neighboring States, and a hatred deep rooted and lasting of all we have been taught to venerate, whether British, Constitutional, or Monarchical, or cling to, in our connection with the Parent State.

Source: English Copyrights Act: Report of the Select Committee,
Canada (Province), Legislative Assembly, 1843, in Parker,
The Beginnings of the Book Trade in Canada, 110–11.

Canadian publishers' ability to ground their business in reprinting—which would be, indeed, the only way they could ever afford to publish Canadian authors. Another major problem for the Canadian book trade was the British publishers' habit of selling North American rights to U.S. publishers, thus shutting out the possibility of a Canadian edition—or of flooding the Canadian market themselves with cheap "colonial editions." In 1889 Canada passed a Copyright Act that required books and periodicals to be manufactured in Canada in order to obtain copyright there—thus in effect removing itself from the Berne Convention. But British power was bluntly applied. Canada passed this bill again in 1890, 1891, and 1895, but every time it was turned back by the Colonial Office. Not until 1899 was approval given for a bill that prohibited importation of foreign-produced editions of books that were already printed in Canada. The 1900 amendment to this legislation became the basis of the agency system of publishing, which ruled the Canadian trade until the 1960s.[28]

Canada's lack of power to develop copyright law suited to its situation clearly hampered the development of its publishing industry in the nineteenth century. In *Dominion and Agency*, a history of the relation between copyright and the emergence of Canadian literature, Eli MacLaren observes that "the structural hindrances to Canadian publishing drove the most ambitious writers to leave home in order to be closer to the people who successfully published books"[29]—a phenomenon captured in the title of Nick Mount's book, *When Canadian Literature Moved to New York*. Canada did not have jurisdiction over its own copyright until it crafted, in 1921, a law that was so close to the U.K. Act of 1911 that it was easily approved in London. It was not until 1982, with the repatriation of the Constitution, that Canada was able to craft copyright law free from the United Kingdom. But by then it was party to the Berne Convention, so in fact there has never been a time when Canada has had autonomy over its own copyright law. And we must insist again that it is not only imperial relations and treaties that colour Canada's copyright situation: Canada has always felt direct effects of U.S. copyright law. The manufacturing clause of the U.S. 1891 Copyright Act created a situation in which any book published independently in

Canada automatically forfeited U.S. copyright. This huge barrier to the development of a Canadian publishing industry was not fully removed until the United States joined the Berne Convention in 1988.[30]

Modern Canadian Copyright

Although representatives of the large cultural industries have been fond of saying that the Canadian Copyright Act is outdated, the 1924 Act was

Table 2. Recent Copyright Reform Bills

Bill	Who and When	What Happened to It
C-60	38th Parliament, 1st session (Liberal minority) Introduced June 2005	Died on order paper when Parliament dissolved November 2005
C-61	39th Parliament, 2nd session (Conservative minority) Introduced June 2008	Died on order paper when Parliament dissolved September 2008
C-32	40th Parliament, 3rd session (Conservative minority) Introduced June 2010 2nd Reading and referral to Special Legislative Committee: November 2010	Died on order paper when Parliament dissolved March 2011
C-11	41st Parliament, 1st session (Conservative majority) Introduced September 2011 (same text as Bill C-32) House 3rd Reading: June 2012 Senate 3rd Reading: June 2012	Royal Assent: June 2012 Proclaimed in partial force November 2012

amended ten times between 1931 and 1997.[31] Throughout the twentieth century, different branches of government produced or commissioned many reports on various aspects of copyright reform.[32] All of these moments have their interest, but we will focus here on twenty-first century developments, which are already historical in their own right. Since 2000, four copyright reform bills have been introduced in the Canadian Parliament. Most recently the enactment of Bill C-11 (the Copyright Modernization Act) in June 2012 has brought a series of major changes to Canadian copyright. As with earlier phases of Canadian copyright history, we have to understand this reform process in both its domestic and its international dimensions.

Until the twenty-first century, copyright was a fairly obscure policy matter. It wasn't clear to most Canadians that it had anything to do with them. But between 2001, when the then-ruling Liberals began consultation on copyright reform, and 2012, when a Conservative majority passed major changes, copyright became a topic of considerable public interest and controversy. Canada had signed two World Intellectual Property Organization copyright treaties in 1996 (the so-called Internet treaties), and the question was how to implement them. The U.S. government, the Department of Canadian Heritage, and the Canadian Recording Industry Association (CRIA, which has since changed its name to Music Canada) in particular were eager to see Canada make circumvention of digital locks illegal, make downloading and file-sharing illegal, and make various other changes to protect or even enhance owners' rights. But partly because of the vicissitudes of the parliamentary system, in which minority governments are vulnerable to defeat at any moment, partly because of the Supreme Court's 2004 ruling on fair dealing in *CCH v. Law Society of Upper Canada* (see chapter 5), and partly because of emerging users' rights activism, it took a decade for Parliament to pass new legislation. Canada thus had an opportunity to see how digital technologies developed and to study the effects of laws that other countries had passed. Whether it took due advantage of this opportunity remains a hot topic of debate, but it is clear that the 2012 bill is much different from proposals entertained ten years earlier.

A few landmarks in this process deserve attention. The first, perhaps, was the Standing Committee on Canadian Heritage's Interim Report on Copyright Reform of 2004, usually known as the Bulte Report after the Chair of this committee, Liberal MP Sarmite Bulte. Developed after considerable public consultation, the report nonetheless presented a very one-sided approach to copyright reform, recommending for example that Internet service providers (ISPs) be required to take down material upon the mere allegation of copyright infringement, and that schools be required to pay licence fees for use of materials available on the Internet. Telecom companies were not pleased and, meanwhile, a new activism was emerging as a growing awareness of consumer rights and public interest in knowledge sharing converged. Bolstered by the *CCH* case, many teachers, students, librarians, and members of the general public became mobilized. One early indicator of a sea change in ideas about copyright was the policy switch of the New Democratic Party that same year. Charlie Angus, a musician and writer, became the party's Culture Critic and promptly declared himself a supporter of users' rights and educational priorities—insisting that artists' interests were distinct from the interests of cultural industries and some-times allied with the interests of teachers and students.[33] In the 2006 elec-tion, the fact that Sarmite Bulte was the beneficiary of a fundraising dinner sponsored by the music industry contributed to her defeat.

When Stephen Harper became prime minister in 2006 with a minority government, the Conservatives were eager to toe the U.S. line with a max-imalist copyright agenda. Their 2008 attempt, Bill C-61, reflected this one-sided approach, generally following the direction of the Bulte Report. But the Conservatives were a bit torn: while on the one hand, they wished to please corporate interests (as they had with rapid passage of a law outlawing camcording in cinemas following Arnold Schwarzenegger's visit to Ottawa in 2007),[34] they understood that their populist support depended on respect for consumers' rights. And those consumers continued to get more vocal and better organized, expressing their views on social media and at con-stituency events. Michael Geist of the Faculty of Law at the University of Ottawa became ever more prominent as a blogger and commentator on the users' rights side.[35] The clout of the content industry and creators lobby was reduced because the Conservatives did not share the Liberals' commitment to cultural nationalism. And the sense of what creators' needs were became more nuanced: the Appropriation Art Coalition, for example, insisted that artists needed a law that permitted established practices of parody and pastiche.[36]

By the time the government tabled Bill C-32 under the leadership of Industry Minister Tony Clement, it had toned down the owners' rights provisions of the reforms, and the agenda started to look more balanced. The resulting Copyright Modernization Act, ultimately passed in June 2012 as Bill C-11, was a compromise. While it retained the restrictive digital locks provisions from Bill C-61, it also extended educational and consumer uses of copyright material. Many in both the cultural and educational sectors are unhappy about various aspects of the bill; the media sector seems more content. Time will tell how effective its balance is.

Part of the international context for this reform process was direct U.S. pressure, through ritual inclusion of Canada in the U.S. Trade Representative's annual Watch List, and no doubt through behind-the-scenes political and diplomatic pressure. But the other international context is treaties. Because of heavy lobbying by the United States in their development, treaties may be seen as U.S. pressure indirectly applied. We have already mentioned the 1996 World Intellectual Property Organization Internet treaties, properly titled the WIPO Copyright Treaty (WCT) and WIPO Performances and Phonograms Treaty (WPPT). WIPO is a descendant of the body that administered the Berne Convention, which harmonized copyright rules for the first time in 1886. Berne sets certain minimum standards, provisions that the signators are expected to comply with. For example, the agreement requires that certain sorts of works be covered and that moral rights be granted; it also states that copyright must exist whether or not a work is registered. But Berne has no effective enforcement mechanism, and recently efforts have been made towards incorporating intellectual property agreements within international trade agreements.

The World Trade Organization (WTO) agreements include an understanding on trade-related aspects of intellectual property (TRIPS). The

TRIPS agreement contains standards for the protection of intellectual property, including copyright, that go well beyond the standards of the Berne Convention. The agreement is enforceable as part of the overall apparatus of the WTO and is subject to the same dispute settlement provisions as other trade agreements, such as the General Agreement on Tariffs and Trade (GATT) and General Agreement on Trade in Services (GATS).

In addition to the WTO framework, various other treaties and agreements also include sections on intellectual property.[37] The North American Free Trade Agreement (NAFTA) contains a chapter on intellectual property, although its provisions are generally similar to those found in TRIPS. The Anti-Counterfeiting Trade Agreement (ACTA) and the Trans-Pacific Partnership (TPP) represent the current generation of international treaties that appear by their names to have little concern with copyright. However, agreements such as these can have major impact on the parameters and shapes of Canadian copyright law, and their development in secret negotiations makes it difficult to challenge them.

PART II

LAW

3. COPYRIGHT'S SCOPE

Copyright is only one of the "big four" intellectual property (IP) devices. The others are patent, trademark, and the law of confidential information and trade secrets.[1] Patent law protects inventions, and trademark law provides a system of avoiding consumer confusion by protecting logos, brand names, and other identifiers used in the course of trade. Trade secrets law protects information closely held within an organization from damaging unauthorized disclosure. While the subject matter of these regimes may sometimes overlap with that of copyright—for example, a trademarked image might also be protected by copyright, and a piece of software might represent part of a patentable device and a copyrighted sequence of code—the principles, laws, and regulatory arrangements of each form of intellectual property are distinct (see Table 3). All of these types of IP have been generating lively and even impassioned

Table 3. Major Types of Intellectual Property

	Copyright	Patent	Registered Trademark	Confidential Information
purpose of rights	to protect forms of expression (does not protect ideas or facts)	to protect inventions (does not protect ideas, algorithms, or scientific theorems)	to protect distinguishing marks from use by others that would create consumer confusion	to protect against unauthorized disclosure of information held as confidential
basis for law	Copyright Act [R.S.C. 1985, c. C-42] (purely statutory)	Patent Act [R.S.C. 1985, c. P-4] (purely statutory)	Trade-marks Act [R.S.C. 1985, c. T-13] + common law	no statute; based on common-law precedents
types of interests protected	literary, dramatic, musical, and artistic works; performers' performances; sound recordings; broadcast signals	inventions—meaning any new and useful art, process, machine, manufacture, or composition of matter, or any new and useful improvement in any art, process, machine, manufacture, or composition of matter	marks used for the purpose of distinguishing certain wares or services from others; certification marks; and distinguishing guise	information held within a firm
requirement for creation of interest	fixation of an original expression in a tangible medium (no formalities required)	application and examination	application and examination	no formalities
term of protection	general rule for works is life of author plus 50 years	20 years from date of filing	15 years from date of registration; can be renewed over and over	as long as the information remains confidential
maintenance during term	protection lasts whether or not the copyright is defended	protection lasts whether or not the patent is defended	can be lost through non-use, non-defence, or failure to renew registration	can be lost through disclosure or failure to take reasonable measures to protect from disclosure

debate around the world in recent years—debates analogous to those provoked by copyright.[2]

But just what does copyright law cover? Chapter 4 will consider which rights copyright owners hold, and chapter 5 will examine how these rights are modified and limited by the rights of users. But before we get there, we need to understand what types of materials can be the subject of copyright protections in the first place, what the requirements are for copyright to begin to exist, and how long the copyright interest lasts.

Copyright's umbrella covers only certain things:
- Copyright subsists in works and other subject matter.
- For copyright to subsist in a work, an original expression must be fixed in some tangible form.
- Copyright applies to original expressions, not to facts or ideas.
- Formalities are not required for a copyright interest to arise; the interest exists at the moment of fixation in a tangible medium of expression.
- Copyright interests are limited in duration, and at the end of the copyright term the materials enter the public domain.

Copyright Subsists in Works and Other Subject Matter

The first copyright act, the Statute of Anne, applied only to particular types of literary works. But as William Hayhurst observes, "During the eighteenth and nineteenth centuries in England, engravers, textile designers, sculptors, dramatists, music publishers, artists and photographers managed to have a succession of statutes enacted. . . . Added to rights to prevent copying were rights to prevent unauthorized public performances of dramatic and musical works."[3] As new representational technologies and cultural practices developed, the category of "works" continued to expand. The current Canadian Copyright Act recognizes four different categories of works—literary, dramatic, musical, and artistic—and they are all very inclusive. Furthermore, since 1997 copyright applies to performers' performances, sound recordings, and broadcast signals. This "other subject matter" carries with it slightly

Table 4. Works and Other Subject Matter Covered by Copyright

Category	Definition
	WORKS
every original literary, dramatic, musical, and artistic work	includes every original production in the literary, scientific, or artistic domain, whatever may be the mode or form of its expression, such as compilations, books, pamphlets and other writings, lectures, dramatic or dramatico-musical works, musical works, translations, illustrations, sketches, and plastic works relative to geography, topography, architecture, or science
artistic work	includes paintings, drawings, maps, charts, plans, photographs, engravings, sculptures, works of artistic craftsmanship, architectural works, and compilations of artistic works
architectural work	any building or structure or any model of a building or structure
book	a volume or a part or division of a volume, in printed form, but does not include (a) a pamphlet, (b) a news-paper, review, magazine, or other periodical, (c) a map, chart, plan, or sheet music where the map, chart, plan, or sheet music is separately published, and (d) an instruction or repair manual that accompanies a product or that is supplied as an accessory to a service
choreographic work	includes any work of choreography, whether or not it has any storyline
cinematographic work	includes any work expressed by any process analogous to cinematography, whether or not accompanied by a soundtrack
collective work	(a) an encyclopedia, dictionary, yearbook, or similar work, (b) a newspaper, review, magazine, or similar periodical, and (c) any work written in distinct parts by different authors, or in which works or parts of works of different authors are incorporated
compilation	(a) a work resulting from the selection or arrangement of literary, dramatic, musical, or artistic works or of parts thereof, or (b) a work resulting from the selection or arrangement of data
computer program	a set of instructions or statements, expressed, fixed, embodied, or stored in any manner, that is to be used directly or indirectly in a computer in order to bring about a specific result
dramatic work	includes (a) any piece for recitation, choreographic work, or mime, the scenic arrangement or acting form of which is fixed in writing or otherwise, (b) any cinematographic work, and (c) any compilation of dramatic works

engravings	includes etchings, lithographs, woodcuts, prints, and other similar works, not being photographs
lecture	includes address, speech, and sermon
literary work	includes tables, computer programs, and compilations of literary works
musical work	any work of music or musical composition, with or without words, and includes any compilation thereof
performance	any acoustic or visual representation of a work, performer's performance, sound recording, or communication signal, including a representation made by means of any mechanical instrument, radio receiving set, or television receiving set
photograph	includes photo-lithograph and any work expressed by any process analogous to photography
plate	includes (a) any stereotype or other plate, stone, block, mould, matrix, transfer, or negative used or intended to be used for printing or reproducing copies of any work, and (b) any matrix or other appliance used or intended to be used for making or reproducing sound recordings, performer's performances, or communication signals
sculpture	includes a cast or model
work of joint authorship	a work produced by the collaboration of two or more authors in which the contribution of one author is not distinct from the contribution of the other author or authors

NON-TRADITIONAL SUBJECT MATTER

communication signal	radio waves transmitted through space without any artificial guide, for reception by the public
performer's performance	any of the following when done by a performer: (a) a performance of an artistic work, dramatic work, or musical work, whether or not the work was previously fixed in any material form, and whether or not the work's term of copyright protection under the Act has expired; (b) a recitation or reading of a literary work, whether or not the work's term of copyright protection under the Act has expired; or (c) an improvisation of a dramatic work, musical work, or literary work, whether or not the improvised work is based on a pre-existing work
sound recording	a recording, fixed in any material form, consisting of sounds, whether or not of a performance of a work, but excludes any soundtrack of a cinematographic work where it accompanies the cinematographic work

Source: Adapted from the Copyright Act.

different constellations of rights than what the Act deems works; rights in non-traditional subject matter (see Table 4) are often known as "neighbouring rights."

Many objects containing intellectual property actually comprise several distinct works. For example, a poetry anthology is a "compilation" in which every poem is also a distinct work with its own particular copyright status. Similarly, a photograph of a sculpture is itself a work, but use of the photograph would in some cases require permission of the owner of copyright in the sculpture. Disputes sometimes arise from this layering of interests. The *Robertson v. Thomson* case of 2006 split the Supreme Court over the issue of the relation between the newspaper publisher's ownership of copyright in the compilation and the freelance writer's ownership of copyright in the individual article.

For Copyright to Subsist in a Work, an Original Expression Must Be Fixed in Some Tangible Form

Originality

A work must be original in order to gain copyright protection. But "original" is obviously a very slippery term: it can mean everything from truly novel (never seen before in all human history) to a much more pedestrian not expressly copied.

For many years the legal test for originality in Canada was that "for a work to be original it must originate from the author; it must be the product of his labour and skill and it must be the expression of his thoughts."[4] Courts had some difficulty in arriving at consistent conclusions based on these principles. In *B.C. Jockey Club v. Standen* (1986), a court held that the compilation of information in horse-racing forms could be protected. In *CCH v. Law Society of Upper Canada* (1999), the trial court held that headnotes summarizing reported court cases lacked a sufficient amount of imagination or "creative spark" to satisfy the originality requirement.[5] Some cases set the bar too low, allowing copyright for works of questionable originality that didn't amount to much more than the laborious collection of data, while others seemed to be setting it too high.

In *Tele-Direct v. American Business Information* (1998), the Federal Court of Appeals tried to articulate a new standard in a case involving a telephone directory—a type of compilation that combines elements that are clearly not under copyright (the raw listing data) and some that clearly are (the layout and presentation of the directory as a whole). The court sought a middle ground: "For a compilation to be original, it must be a work that was independently created by the author and which displays at least a minimal degree of skill, judgment and labour in its overall selection or arrangement. The threshold is low, but it does exist."[6] But this pronouncement, too, was subject to various interpretations. What exactly is meant by "skill, judgment and labour"? Can the requirement be met by an abundance of one or two of the three, or is some of each required?

In 2004 the Supreme Court provided needed clarification in *CCH v. Law Society of Upper Canada*, the case in which the trial court had set a standard of creative spark for originality. The Supreme Court followed the Appeals Court in rejecting that approach. "For a work to be 'original,'" it stated, "it must be more than a mere copy of another work. At the same time, it need

---

Why does it matter whether the standard for originality is too high or too low?

If the standard is too low, facts or ideas could become effectively unavailable for others to use freely or reuse. Copyright extended to an alphabetical directory listing or the collection of publicly available data into an obvious chart or table would effectively confer ownership on the data itself.

If the standard is too high, courts find themselves in the role of evaluating artistic merit, which is not appropriate. Also, works that as a practical matter ought to have copyright protection—pamphlets and instruction manuals, for example—might not meet the requirement.

not be creative, in the sense of being novel or unique" (para. 16). While noting that "creative works will by definition be 'original' and covered by copyright"—in other words, "creativity is not required to make a work 'original'"—Chief Justice McLachlin went on to provide a more precise articulation of the originality test, which is now the standard:

> What is required to attract copyright protection in the expression of an idea is an exercise of skill and judgment. By skill, I mean the use of one's knowledge, developed aptitude or practised ability in producing the work. By judgment, I mean the use of one's capacity for discernment or ability to form an opinion or evaluation by comparing different possible options in producing the work.[7]

The court further noted that the "exercise of skill and judgment will necessarily involve intellectual effort" and "the exercise of skill and judgment required to produce the work must not be so trivial that it could be characterized as a purely mechanical exercise" (para. 25). In this statement, the court explicitly rejected the sweat-of-the-brow doctrine, which accords originality to works simply by virtue of the labour that went into them.

Fixation

Copyright only subsists in works fixed in some tangible form. Interestingly, there is no fixation requirement on the face of the Canadian statute.[8] The fixation requirement comes from case law, and it is related to the principle that ideas are not covered by copyright: according to what is known as the "idea-expression dichotomy," only the embodiment of the ideas is protected.

But where is the boundary between idea and expression? What constitutes fixation? The most often-cited Canadian case about the fixation requirement is *Canadian Admiral v. Rediffusion* (1954), involving televised Montreal Alouettes football games. The gist of the case, at least insofar as fixation is concerned, was that a live broadcast of a game did not meet the fixation requirement. The court contrasted simultaneous, unedited transmission with the broadcast of a taped game, which it did recognize as a fixation. (Since this case, communication signals have been added to the

Many scholars have criticized copyright law's originality requirement, arguing that it implies an overly individualistic, Romantic idea of authorship. Surely, they argue, originality is impossible, given that we all work within shared traditions and languages. Many years ago, Northrop Frye blamed copyright for readers' tendency to exalt an author's contribution over the rich tradition from which it sprang. Mark Rose sums copyright up as "an institution built on intellectual quicksand: the essentially religious concept of originality, the notion that certain extraordinary beings called authors conjure works out of thin air"—whereas, as Jessica Litman puts it, "the very act of authorship in any medium is more akin to translation and recombination than it is to creating Aphrodite from the foam of the sea."

In Canada as elsewhere, though, the courts have not generally understood originality to mean novelty in the strict sense. The law is concerned with which "head" produced the work, and thus refuses copyright to a copied work. But it does not require that the work has no antecedents: it does not demand that a work should "rise spontaneously from the vital root of genius," as the Romantic poets would see it. If it did—if authors had to *prove* originality—copyright law could not function at all. We tend to agree with Paul Saint-Amour, who argues: "The phenomenon of 'copyright creep,' however much one might regret its reapportioning of public and private domains, appears to have resulted from the influence of the private sector on the legislative climate, rather than from some privatizing drive inherent in copyright's metaphysics."

Sources: Frye, *Anatomy of Criticism*, 96; Rose, *Authors and Owners*, 142; Litman, "The Public Domain," 966; Young, "Conjectures on Original Composition," 11; Saint-Amour, *The Copywrights*, 6.

Act as a category in which copyright can subsist—even though they are not fixed—but the case still offers guidance on the nature of fixation.)

In his book *Copyright Law*, David Vaver takes a sceptical view of fixation as a requirement for copyright, arguing that it is better thought of as a rule of convenience than as a fundamental principle.[9] Indeed, the requirement is especially fraught in the digital age, when the distinction between fixation and lack of fixation seems arbitrary. It also seems as if some forms of contemporary art in which the work's essence lies in its changeability—with media such as melting ice, shifting light, or even, notoriously, rotting meat—might have difficulty meeting the fixation requirement.[10] Body art, too, seems to pose problems, as while the copyright would likely belong to the tattoo artist who "fixed" the work to a human body, human rights and freedom of expression laws would prevent the artist from exercising copyright in all its dimensions.

An older challenge to the fixation principle comes in the case of an oral presentation. If a speaker is speaking from notes, clearly the notes are fixed works in which copyright subsists. But what if the words exist only in spoken form? The Copyright Act contains a special definition for "lecture"

Is a joke covered by copyright?

It depends. A good line or two, spontaneously if aptly delivered, would probably not be covered. In *Glenn Gould Estate v. Stoddart* (1998), the court found that the interviewer who elicited Gould's unscripted words owned the copyright in them. A more prolonged piece in a public venue, even if improvised, would count as a performer's performance under section 15 of the Act—according to which the performer alone has the right to fix the performance in any form or to authorize such fixation. But if comics are working from someone else's written material rather than generating their own material in performance, they might need permission, because that material, like all written matter, is born copyrighted.

that includes an address, speech, or sermon (see Table 4, and also the exemption for lecture reporting in Table 8 on pages 84–86)—a definition that seems to constitute an exception to the requirement of fixation. But not all oral presentations are addresses, speeches, or sermons, and this has left storytellers, for example, vulnerable to having their work appropriated by listeners bearing recording devices or notepads. The relatively new category of "performer's performance" would cover many oral presentations. In the case of a performer's performance, the Copyright Act is explicit that there is no requirement for fixation.[11]

Copyright Applies to Original Expressions, Not to Facts or Ideas

Copyright does not apply to ideas or facts; it applies to the way they are presented. This is the crucial distinction of the idea-expression dichotomy. If we allowed parties to claim ownership rights over facts or ideas, cultural innovators would be constrained by lack of access to raw materials. Also, of course, ownership of facts and ideas would have various deleterious effects on democracy and public discourse. Copyright's recognition of the right to an exclusive claim to a particular expression of an idea or fact is a sort of compromise, permitting both reward and foundation for innovation.

Weather reports in a newspaper, for example, might be arranged in a table with the date in one column, the city in the next, and the temperature in the next. No one can own these facts, even if someone puts expertise and expense into collecting them. But a more original presentation format might attract copyright: an animated artistic map, for instance, with moving clouds and a distinctively dressed announcer reading a script that includes the weather forecast. Although a third party can still use the facts presented therein without permission, a full studio production is a different matter from the bland weather table, and it crosses the line from unprotected facts to copyrightable subject matter.

The location of the idea-expression line is often difficult to pinpoint (many of the major thinkers of the twentieth century would say the distinction between ideas and expressions does not exist),[12] but the courts have developed principles for doing so. In *Cuisenaire v. South West Imports* (1969), the plaintiff attempted to apply copyright to a set of coloured rods used for teaching math to young children, as described in a book he had published. The court rejected this claim, noting that copyright applied to the book but not to the rods themselves—the idea embodied in the book.[13]

Where is the boundary between idea and expression in a fictional work? In their abstract form, dramatic plots are ideas, not expressions. For example, the boy-meets-girl-and-their-families-are-outraged plot is not in itself copyrightable. But fill things out a bit with character names, specific wording, and plot twists, and an author may be able to claim copyright. In *Anne of Green Gables Licensing Authority v. Avonlea Traditions* (2000), the defendant, a manufacturer of Anne souvenirs, relied on *Cuisenaire* for the proposition that copyright should not extend beyond a book to cover three-dimensional objects described therein. But the court held that a literary work includes "any of its characters whose descriptions are distinctive, thorough, and complete." It rejected the defendant's parallel with *Cuisenaire*, stating that the Anne merchandise required licensing as an expression of the "detailed verbal portrait" in the literary work.[14] So here we have, in a sense, copyright upheld in an idea—the idea of the characters in a book. (Since Anne has been in the public domain since 1993, her image is now no longer controlled, to the extent that she has even recently been portrayed on a book cover as a bedroom-eyed blonde.)

Courts have sometimes recognized that dangers to innovation can arise from understanding "expression" too expansively. In a dispute between two computer software companies, *Delrina v. Triolet Systems* (2002), the court noted, "If there is only one or a very limited number of ways to achieve a particular result in a computer program, to hold that that way or ways are protectable by copyright could give the copyright holder a monopoly on the idea or function itself."[15] This case shows how the idea-expression dichotomy will act as a limit on the scope of copyright in a case in which a particular idea can only be expressed in one way (or in only a very few ways). In this situation, it is said that the idea "merges" with the expression of the idea, and copyright cannot be applied.

The idea-expression dichotomy also provides the basis for withholding copyright protection from the elements in a database. In Europe, special legislation provides statutory protection for databases, and the North American database industry has been clamouring for similar legislation for many years. It seems unlikely that database legislation will have much of a chance in Canada, although it remains on the wish list of some entrepreneurs in the information industry.

Formalities Are Not Required for a Copyright Interest to Arise; the Interest Exists at the Moment of Fixation in a Tangible Medium of Expression

One of the persistent myths about copyright is that in order to exist, a copyright must be registered and the work marked with the copyright symbol ©. In fact, copyright exists at the moment an original expression is fixed into some tangible medium (or, for copyright's other subject matter, at the moment of broadcast, performance, or sound recording). It does not matter legally whether the work is marked with a ©, and it does not matter whether or not the copyright is subsequently registered, at least insofar as the initial question of validity of copyright is concerned.

Hardly a day can go by in the life of most members of modern society without the creation of many works in which copyright subsists: grocery lists, emails, doodles, unfinished love notes, and voice memos with some small modicum of originality are covered automatically by copyright— along with novels, rock operas, and scientific treatises. You probably do not assert copyright in your everyday creations, but you could. In using works created by others, then, you should never assume that you do not have to worry about infringement just because there are no apparent markings. You may be just as guilty of infringement of copyright materials that lack such a marking and that are unregistered.[16]

Even though a copyright mark is not required, you may have reason to affix one, or even to go to the trouble and expense of registering a copyright. Copyright marking puts the whole world on notice of your copyright interest.

How do I register copyright in my work?

Go to www.cipo.ic.gc.ca—the website of the Canadian Intellectual Property Office—and follow the instructions. The process is straightforward and does not require the assistance of a lawyer. As of 2013, the basic registration fee is $50. Your work will be listed in the publicly available database of registered Canadian copyrights.

Short of litigation, it reminds other users that the materials are subject to copyright even though they may be readily accessible in a library or on the Internet. You can mark your work with a © whether or not you have registered it.

Registration goes even further to create a presumption of ownership. If you ever have to defend your copyright in court, registration will prevent an infringer from claiming not to have known that your work was copyrighted.[17] Registration also sets a creation in a time frame that could be important if a priority dispute arises with another party. However, the vast majority of copyright owners never register their copyrights. As the Canadian Intellectual Property Office states on its website:

> Registration is no guarantee against infringement. You have to take legal action on your own if you believe your rights have been violated. Also, registration is no guarantee that your claim of ownership will eventually be recognized as legitimate. Note too, that the Copyright Office does not check to ensure that your work is indeed original, as you claim. Verification of your claim can only be done through a court of law.[18]

A growing number of people simply don't want to assert all the rights that copyright automatically grants them. You might, for example, feel that it's fine if people make non-commercial use of your work; you just want payment if they make money from the reuse. Or you might have no problem with unpaid reuse as long as the work is not changed. If you only want "some rights reserved," you can choose a licence through Creative Commons (http://creativecommons.ca; see chapter 17). This process is free and very straightforward.

Copyright Interests Are Limited in Duration, and at the End of the Copyright Term the Materials Enter the Public Domain

Many people think of copyright as a form of private property, similar to personal goods or real estate. But unlike the ownership interests in personal property, real estate, or even other intangible goods, the duration of a copyright interest is limited in time. In Canada the term of protection for copyright is generally the life of the author plus fifty years—or as section 6 of the Act puts it "The term for which copyright shall subsist shall, except as otherwise expressly provided by this Act, be the life of the author, the remainder of the calendar year in which the author dies, and a period of fifty years following the end of that calendar year." The idea here is that the children and grandchildren of the author or copyright owner have a chance to benefit from the estate. Thus the term of copyright in the work of Paul Kane, who died in February 1871, ended at midnight on 31 December 1921; the term of copyright in the work of Lucy Maude Montgomery, who died in April 1942, ended at midnight on 31 December 1992.

At the end of the copyright period, the copyright interest automatically lapses and the work enters the public domain. All works in which copyright subsists are destined to become part of the public domain; it is just a matter of time. You can think of copyright term as a moving wall between today's creators and a shared heritage: the constant renewal of the public domain ensures that creators have a growing mass of resources with which they can work freely—in both senses of the word.

Different countries have different copyright terms: in the United States, copyright generally lasts for the life of the author plus seventy years, and in Mexico the term is life of the author plus one hundred years. But because

Table 5. Duration of Copyright Term in Special Cases

Subject	Copyright Term	Copyright Act Section
known author	life of author + 50 years	6
unknown author(s)	earliest of (a) publication + 50 years, (b) making + 75 years	6.1, 6.2
known joint authors	life of last surviving author + 50 years	9(1)
works owned by the Crown	publication + 50 years	12
performer's performance	fixation + 50 years, or performance + 50 years if not fixed	23(1)(a)
sound recording	making + 50 years	23(1)(b)
broadcast	broadcast + 50 years	23(1)(c)

copyright is national in application (its jurisdiction is limited to the territory of the country that enacts it), anyone in Canada follows the term stipulated in the Canadian Act, no matter what citizenship they hold or what the origin is of the material in question. This principle includes websites located on Canadian servers. Even though they may reach people elsewhere, they operate under Canadian law. After all, the alternative—governing a Canadian website according to the longest copyright term in the world, rather than by our own laws—is preposterous. Given variations between jurisdictions, though, it might be wise to include a note on the site reminding users that "public domain" is defined with respect to Canadian law and users in other jurisdictions are responsible to know the law that applies there. And once commercially distributed outside of Canada, Canadian-generated materials are bound by the copyright terms of those other jurisdictions.

Within the Canadian Copyright Act, there are special cases for copyright term in certain classes of works or other subject matter (see Table 5). Note that since 2012, the copyright term in photography follows the general rule for works—photography used to have its own specific treatment. Also note that even if the first owner of a work is a corporation (see chapter 7), the copyright term of that work is determined by the date of death of the author, if that author is known.

Unpublished works represent another case of specific copyright term issues. Prior to 1999 the copyright duration for unpublished works was perpetual. If a work was eventually published—with permission of the heirs or estate, if they could even be identified—the term would last for fifty years from that point. Since many important works (such as manuscripts, works in progress, and private correspondence) remain unpublished during the life of an author, perpetual copyright created serious problems for historians and other scholars. The 1997 amendments to the Act improved this situation, but they are bafflingly complex, and furthermore they created a lamentable and anomalous waiting period for archival materials whose authors died between 1949 and 1998. In these cases, strangely, the authors' published works will enter the public domain before their unpublished works (see Table 6).[19]

Table 6. Duration of Copyright for Unpublished or Posthumously Published Works

Category of Work	Date of Expiration of Copyright
Work published after author's death but before 1 January 1999	End of calendar year of publication + 50 years
Unpublished work of author who died before 1 January 1949	31 December 2004
Unpublished work of author who died between 1 January 1949 and 31 December 1998	31 December 2049
Unpublished work of author who died after 31 December 1998	Year of author's death + 50 years

Every January 1, blogger Wallace McLean celebrates Public Domain Day by listing authors, composers, architects, and other creators whose work has entered the Canadian public domain. On 1 January 2013, he inducted into "common cultural property" a bumper crop of luminaries including William Faulkner, Isak Dinesen, Eleanor Roosevelt, Vita Sackville-West, Hermann Hesse, e.e. cummings—not to mention Ludwig Bemelmans (the author of the Madeline books) and Irma Rombauer (the author of *The Joy of Cooking*).

McLean notes that other jurisdictions and unpublished materials have longer terms; he ends his 2013 entry with the declaration, "Short live copyright, and long live the public domain!"

Source: Wallace McLean, "Public Domain Day 2013," http://publicdomain.xanga.com.

At the age of 18, out of his home in Burnaby, British Columbia, musician Edward Guo started the International Music Score Library Project (http://imslp.org), a virtual library of public domain music scores. Running as a wiki to which others can contribute, it is now a major and growing resource for classical musicians, with 221,000 scores and 250,000 downloads per day. The path didn't always run smoothly for Guo, however; in 2007 he received a takedown notice from a Vienna publisher, Universal Editions, followed by another from a Canadian lawyer they had retained. It was a trial by fire:

> I was 19 years old at the time, and the week after I received the second letter was probably the worst week of my life up to then. I had absolutely no idea what to do, and I remember we were discussing possibilities of jail time and so forth. . . . In the end I felt the best course of action was to take down the entire site, regroup, and go online again. That turned out to be the correct decision. The amount of support for IMSLP and the amount of negative PR for UE helped IMSLP to go back online nine months afterwards. We have not heard anything from UE since. We did, however, receive other takedown notices after we went back online (though these also seem to have slowed down—we haven't gotten one in 6 months I believe). After a while you start getting used to all the fantastical claims of copyright on works that were composed 400 years ago (we got one for a reprint of an old edition of Pachelbel's Canon), and it is pretty much just business as usual. One of my biggest gripes with the [U.S.] DMCA takedown system is the lack of penalties for frivolous, or even blatantly false, takedown notice. . . . Copyright owners are very well protected under copyright law; why shouldn't the public domain be as well protected?

Mr. Guo's story is an inspiring one—not only did he maintain and expand this essential resource for musicians, but he is now a graduate of Harvard Law School!

Sources: http://imslp.org, email from Edward Guo, 9 December 2012.

4. OWNERS' RIGHTS

Assuming that all of the conditions for copyright to subsist have been met, the first owner has not *one* copyright but rather a series of distinct rights—or in the usual copyright metaphor, a bunch of sticks in a bundle. Chapter 7 will explore the question of who the owner of a copyright is, but for now you can think of the first owner as a creator, or a company or institution under whose auspices a work was created. Each of the copyrights is enumerated in the Copyright Act. Each confers the ability to do some specific thing with the material in question, and also to exclude other people from doing so without permission. The economic rights can be separately deployed: the first owner might keep some, sell or license others to distinct purchasers under particular conditions, and give one or two away. Authors' moral rights, in contrast, can be waived, or bequeathed upon death, but not assigned to others.

The key section of the Copyright Act for purposes of determining the economic rights of the owner of a work is section 3(1), which sets out three core rights in its preamble, and lists ten specific instances of these rights in paragraphs (a) through (j), adding at the end the right to authorize any of these instances. While each of those rights is different, and many apply only to certain types of works, in every instance the right is a "sole right." This means that the owner of the right not only can do the thing specified, but also can exclude others from doing it. For example, the right of reproduction does not just grant the owner the right to make reproductions of the work, it excludes the rest of the world from doing so without the owner's consent. The word "sole" is why copyright is referred to as a sort of monopoly.

Keep in mind, too, that each of these rights is severable and freely assignable on its own. People often use the term "derivative rights" to refer to owners' rights to control follow-on use of their work, but this umbrella term does not appear in the Canadian Act, which instead specifies each right individually.[1]

The Reproduction Right

Section 3(1) starts with the "sole right to produce or reproduce the work or any substantial part thereof in any material form whatever." In questions of copyright, the reproduction right is usually the first thing that comes to

Table 7. Rights in Works and Other Subject Matter

Subject Matter	Pertinent Section of the Act
works	3(1)
performer's performance	15(1)
sound recording	18(1)
communication signal	21(1)

For the purposes of this Act, "copyright," in relation to a work, means the sole right to produce or reproduce the work or any substantial part thereof in any material form whatever, to perform the work or any substantial part thereof in public or, if the work is unpublished, to publish the work or any substantial part thereof, and includes the sole right

(a) to produce, reproduce, perform or publish any translation of the work,

(b) in the case of a dramatic work, to convert it into a novel or other non-dramatic work,

(c) in the case of a novel or other non-dramatic work, or of an artistic work, to convert it into a dramatic work, by way of performance in public or otherwise,

(d) in the case of a literary, dramatic or musical work, to make any sound recording, cinematograph film or other contrivance by means of which the work may be mechanically reproduced or performed,

(e) in the case of any literary, dramatic, musical or artistic work, to reproduce, adapt and publicly present the work as a cinematographic work,

(f) in the case of any literary, dramatic, musical or artistic work, to communicate the work to the public by telecommunication,

(g) to present at a public exhibition, for a purpose other than sale or hire, an artistic work created after June 7, 1988, other than a map, chart or plan,

(h) in the case of a computer program that can be reproduced in the ordinary course of its use, other than by a reproduction during its execution in conjunction with a machine, device or computer, to rent out the computer program,

(i) in the case of a musical work, to rent out a sound recording in which the work is embodied, and

(j) in the case of a work that is in the form of a tangible object, to sell or otherwise transfer ownership of the tangible object, as long as that ownership has never previously been transferred in or outside Canada with the authorization of the copyright owner, and to authorize any such acts.

— Copyright Act, section 3(1).

mind. When you use a photocopy machine or a scanner, copy and paste on your computer, or make a digital copy of a CD, you are implicating the reproduction right. (By using the word "implicating," we mean that you are in a position where you ought to ask whether you are infringing—though as we will see, you can implicate a right without infringing it.)

What exactly does it mean to "produce or reproduce" a work? Or to put it another way, what activity or conduct is required to trigger the operation of this section? Usually, the situation is straightforward. If you put a book on a photocopy machine and press Start, you are implicating the reproduction right.

But sometimes the border between what is or is not a production or reproduction is less clear. The 2002 Supreme Court case *Théberge v. Galerie d'Art du Petit Champlain* wrestled with the copyright implications of a chemical process that can remove an image from a piece of paper and transfer it to a canvas. In a split decision, the majority of the judges decided that transferring the image from one medium to another did not constitute a reproduction. For them, the important question was whether the number of works increased: at both the start and end of the process, there was one work. For the minority, the important point was that there was a new and independent fixation of the work; even though the source was destroyed, a new work was "produced."[2]

A more common source of controversy is the substantiality requirement. The Act says that the owner of the reproduction right has an exclusive right to reproduce the whole work "or any substantial part thereof." This means that the owner's right to exclude does not extend to reproductions that are *not* substantial. The Act, however, does not define what is meant by "substantial." Where is the line between a non-substantial snippet and a substantial reproduction that implicates the reproduction right? Given the indistinctness of this threshold, many potential users are unsure of what is permissible, and rights holders have been using intimidation tactics to assert their preferred norm for substantiality. The key here is that in determining whether the threshold of substantiality has been met, courts have looked to both quantitative and qualitative factors, and they have not laid out any magic number. If anyone insists on using a rule of thumb for substantiality of 5 per cent or three seconds or two column inches—or any such handy formula—they are simply wrong, as *Warman v. Fournier* (2012) makes very clear.[3]

Warman v. Fournier arose out of a dispute between the operators of a conservative-leaning website, Mark and Constance Fournier, and a well-known human rights lawyer, Richard Warman, about materials that the Fourniers posted to their website www.freedominion.ca. The Fourniers had posted a *National Post* article written by Jonathan Kay under the title "Jonathan Kay on Richard Warman and Canada's Phony-Racism Industry." After a demand to remove the article, the Fourniers replaced it with excerpts. Warman, who had obtained the rights to the article, alleged that these excerpts were still substantial reproductions of the work within the meaning of section 3(1) of the Act.

In a significant passage analyzing the meaning of the substantiality requirement in section 3, the court found that the requisite substantiality was lacking and that there was no infringement. In determining substantiality, the court applied a five-part test:

> The relevant factors to be considered include:
> a. the quality and quantity of the material taken;
> b. the extent to which the respondent's use adversely affects the applicant's activities and diminishes the value of the applicant's copyright;
> c. whether the material taken is the proper subject-matter of a copyright;
> d. whether the respondent intentionally appropriated the applicant's work to save time and effort; and
> e. whether the material taken is used in the same or a similar fashion as the applicant's.

As to the first factor, the court said:

Quantitatively, the reproduction constitutes less than half of the work. The Kay Work itself consists of a headline and eleven paragraphs. The reproduction on Free Dominion included the headline, three complete paragraphs and part of a fourth. Qualitatively, the portions reproduced are the opening "hook" of the article, and the summary of the facts on which the article was based. Most of the commentary and original thought expressed by the author is not reproduced. (para. 25)

Concerning the fourth factor, the court stated:

It does not appear that the excerpts of the Kay Work were reproduced to "save time and effort." Based on the context of the posting, the respondents reproduced portions of the Kay Work to preserve a record of the facts summarized in the article, so that members of Free Dominion could continue to discuss those facts on the forum. Also, contrary to the applicant's argument, the reproduction does include a summary or paraphrase of part of the work, specifically the second paragraph. (para. 27)

The court did not find the other substantiality factors particularly relevant to this dispute and concluded:

. . . considering the matter as a whole I find as a fact that the applicant has not established that the excerpts of the Kay Work reproduced by the respondent constitute a "substantial part" of the work, and there is therefore no infringement. (para. 28)

The court continued its analysis to find that "even if the reproduced portions of the Kay Work amount to a substantial part, I find that the respondents' reproduction constitutes fair dealing for the purposes of news reporting, pursuant to section 29.2 of the *Copyright Act*" (para. 29).

We will return to fair dealing in chapter 5, but with regard to substantiality, *Warman v. Fournier* confirms that the bar may be set quite high,

How long a quotation counts as "substantial" use requiring permission?

Apply the *Warman v. Fournier* tests to find out. No one test is decisive: it's the overall weight of the answers that counts. For example, are you quoting because you're too lazy or cheap to write your own copy (test d)? If so, you already know the right thing to do: ask permission. Are you quoting most of the work, more than you need to, and/or a work's most striking parts (test a)? That would weigh in the direction of permission, too. But even then, there may be mitigating factors as you work through the tests as a group. Quotation can actually add to the value of someone's copyright by enhancement of reputation or publicity (test b), or it may simply take the material into a different realm and thus not compete with it at all (test e). If quoting is part of the normal practice of your community, as it is, for example, in scholarly discussion or reporting, a court would likely give weight to that, as the Supreme Court did in *CCH v. Law Society of Upper Canada* concerning the "character of the use" test for fair dealing (see chapter 5). Think of the substantiality requirement as a safety valve to ensure that copyright doesn't become too much of a drag on the vitality of public discourse. As we discuss in chapter 17, quoting is the essence of democratic dialogue and cultural growth: it allows us to knit our ideas together, to leverage new ideas, and to demonstrate credibility. An expectation of permission for the reproduction of every tweet and sound bite would slow down discussion, and sometimes smother it. Through the substantiality requirement, the Copyright Act and the courts have given us permission to not always get permission.

contrary to the claims of many rights holders. The court essentially said that "substantial" really means *substantial*: permission for anything less is simply not necessary.

The Public Performance Right

The second core right, also contained in the first sentence of section 3(1), is "the sole right . . . to perform the work or any substantial part thereof in public." But what does it mean to "perform the work," and when is that performance deemed to be "in public"?

The Copyright Act defines "performance" as "any acoustic or visual representation of a work . . . including a representation made by means of any mechanical instrument, radio receiving set or television receiving set." You can perform a work by singing it or acting it out—but also by playing a performer's performance of it on an MP3 player or other device, because the work is contained within the performance of the work. The Act uses the clumsy term "performer's performance" to differentiate a performance by a human from a performance by a radio, television, or other technology.

The leading Canadian case discussing the public performance right is *Canadian Admiral v. Rediffusion* (1954)—the same case cited earlier on questions of fixation (see chapter 3). In this case, revolving around televised football games, the court concluded that the rebroadcast of the games was not a matter of public performance because it was done in the private homes of cable subscribers. The court rejected the argument that a large number of private performances, because of the number, become public performances. It reasoned that the character of the individual audiences remained the same, each being private and domestic and therefore not "public."[4]

Showing a work in a public theatre where admission is being charged is clearly an instance of a public performance, and viewing a video in the privacy of your home is clearly not. But in between these two extremes are all sorts of grey areas. One example is the classroom setting, where it is understood that only a number of students registered for a course will be in the room to see the performance of the work. Passersby would not be invited in. Is this more analogous to the public theatre or the private home? The courts have never been asked—but in the meantime, the Copyright Modernization Act of 2012 (Bill C-11) created an exception to permit classroom screenings and render this particular question moot.[5]

The First Publication Right

The opening paragraph of section 3(1) concludes with the first publication right. In the case of a work that is unpublished, the owner has "the sole right . . . to publish the work or any substantial part thereof." This simply means that the owner has the right to publish a work for the first time—or to withhold it from publication.

Unlike other section 3 rights, which are persistent, the first publication right is exhausted upon publication. The Act provides that a work is considered published when copies of the work are made available to the public.[6]

Other Enumerated Rights

After setting out the three core owners' exclusive rights, section 3(1) goes on to list additional rights, which the Supreme Court in *Entertainment Software Association v. SOCAN* (2012) considered examples of or subsidiary to the three core rights in the preamble (para. 42).[7]

Is it copyright infringement to link to someone else's website?

To demonstrate copyright infringement, the copyright owner has to show that the infringer did something that only the owner can legally do, and that there was no permission. Linking to someone else's web page, even against their wishes and without their consent, simply does not implicate any of the copyright owner's rights as they are listed in section 3 of the Copyright Act: it is not reproduction, for example, or distribution, or authorization. It might violate someone's sense of Internet etiquette, and in some cases it might even be considered a breach of a contract, but it is not copyright infringement. (For more detail, see chapter 10.)

The Translation Right

According to section 3(1)(a), the first owner also has the sole right "to produce, reproduce, perform or publish any translation of the work." It has
long been understood that a translation of a work is an original literary work
of which the translator is the author. However, the rights of the original
author stand: translation requires permission.

In recent years, computer code has produced the main question about
translation rights. Is it a "translation" within the meaning of the translation
right to rewrite a computer program in a different programming language?
In *Apple Computer v. Mackintosh Computer* (1986) the federal trial court held
that conversion from one code to another did not constitute a translation.[8]
But in another case, *Prism Hospital Software v. Hospital Medical Records Institute* (1994), a B.C. court ruled that the conversion of a program into a different programming language did constitute a translation. There still seems to
be quite a bit of uncertainty in this area.

The Conversion Rights

Although works are categorized as literary, dramatic, musical, or artistic, a
work often starts out in one category and is later adapted into another. Thus
a poem becomes a song, a movie becomes a novel, or a novel becomes a play.

In the nineteenth century, authors did not have translation rights. When
Harriet Beecher Stowe went to court in 1853 over an unauthorized translation of *Uncle Tom's Cabin* into German, the court said that while she retained
the "exclusive right to print, reprint and vend . . . 'copies of her book,'" a
translation was to be considered "a transcript or copy of her thoughts or
conceptions, but in no correct sense . . . a copy of her book."

Source: *Stowe v. Thomas.*

In each case, the adapter would require a licence because such "conversions" are the sole right of the owner under sections 3(1)(b) through (e).[9]

The Public Communication Right

Of increasing importance in the networked environment is section 3(1) (f), which sets forth the right "to communicate the work to the public by telecommunication." While this right once mainly regulated the behaviour of broadcasters, in the era of computer networks it potentially affects many more people. In *SOCAN v. Canadian Association of Internet Providers (CAIP)* (2004) the Supreme Court was called on to test the communication right in the context of Internet circulation. The court concluded that Internet service providers do not themselves actively communicate to the public material circulated through their servers and do not infringe copyright by functioning as a conduit. And more recently, in *ESA v. SOCAN* (2012), the Supreme Court has made it clear that the scope of the communication right is limited: it is not cumulative with the reproduction right. In other words, the same activity cannot implicate both the reproduction right and the public communication right, as this could result in double compensation. In the context of communications over the Internet, the court held that a download constitutes a reproduction, but it does not also then constitute a communication to the public by telecommunication. A stream, on the other hand, constitutes a communication to the public but not a reproduction.

The Exhibition Right

Section 3(1)(g) stipulates a "public exhibition" right for certain artistic works created after June 7, 1988. This means that even if a gallery or an individual owns a work of art, a fee must be paid upon exhibition for as long as its copyright subsists and as long as the work is not being exhibited with the aim of selling or renting it. The artist may negotiate that fee individually or through a collective such as Canadian Artists' Representation Copyright Collective (CARCC). We discuss this right further in chapter 12 on Visual Arts.

SOCAN v. CAIP (2004)

The situation:

The Society of Composers, Authors and Music Publishers of Canada (SOCAN) collects tariffs for the performance or communication to the public of music. Its Tariff 1, for example, is for music played on the radio, and its Tariff 3 is for cabarets, cocktail lounges, and bars. In 1999 SOCAN proposed Tariff 22 for music performed or communicated to the public via the Internet.

SOCAN argued that a communication to the public occurs when the end-user can gain access to a musical work from a computer connected to a network—and thus that virtually everyone involved in the Internet transmission chain is liable for the communication, including those who provide transmission services, operate equipment or software used for transmissions, provide connectivity, provide hosting services, or post content. The Canadian Association of Internet Providers (CAIP) argued the opposite.

The result:

The Copyright Board rejected SOCAN's position. In other words, while a person who posts music on an open server authorizes its communication to the public, the typical activities of an Internet service provider do not constitute communication by telecommunication to the public. As a result, there would be no liability for an ISP to pay a royalty. Upon appeal, the Copyright Board's decision was mostly upheld at the Federal Court of Appeal, and entirely upheld at the Supreme Court.

The Rental Rights

Generally speaking, the exclusive rights of an owner do not include the right to rent out the work. In *Théberge v. Galerie d'Art du Petit Champlain* (2002), the Supreme Court described an established rule when it noted, "Once an authorized copy of a work is sold to a member of the public, it is generally for the purchaser, not the author, to determine what happens to it." But section 3(1) contains two specific exceptions in which the owner is granted exclusive rental rights. Section 3(1)(h) applies to reproducible computer programs, and section 3(1)(i) applies to sound recordings that embody a musical work. This means that rental of computer programs or sound recordings without the permission of the owner is a violation of one of the owner's rights. But libraries and individuals are allowed to loan these materials without charge.

The Distribution Right

Section 3(1)(j), added to the Act in 2012, provides a first owner's exclusive right "to sell or otherwise transfer" a work in the form of a tangible object. The new right only applies in situations where the ownership of the object has never previously been transferred with the authorization of the copyright owner. Frankly, the meaning of the provision is cryptic.[10] It would seem that if a third party were to take an unpublished work—say, a piece of video— and sell it in physical form without the permission of the owner, that would already be infringing the first publication right. It is not clear to us what extra mileage a rights holder gets from this provision. From the users' rights end, it could perhaps be taken as a reaffirmation of what in the United States is called the "first sale doctrine." This principle holds that the physical container of a work (i.e., a book or a record) must be sold the first time with the owner's authorization, but after that it may be sold freely by anyone.[11]

The Authorization Right

Last among the economic rights, but certainly not least, is the authorization right. Section 3(1) closes with the words "and to authorize any such acts." "Any such acts" refers to all of the rights set out in 3(1), and in essence incorporates the entire section.

Unfortunately, the term "authorization" is nowhere defined in the Act. What does it mean to authorize someone to do something? Telling your assistant to go to the photocopy machine, punch in your copy code, and make a copy of an article for everyone in your workplace is probably author-ization to reproduce a work. But what if you just provide a copy machine or a scanner for others to use as they wish? Or what about just providing a com-puter, which by its nature has ample copying capacities? The overarching question here is where to draw the line between being a significant enabler of a particular result and doing something that is merely part of a causal chain.

In recent years, case law has placed some constraints around the potentially enormous scope of the authorization right. In *CCH v. Law Society of Upper Canada* (2004), the Supreme Court rejected the publishers' line of argument that a library's provision of free-standing photocopiers constituted a violation of publishers' authorization rights. "A person does not authorize infringement by authorizing the mere use of equipment that could be used to infringe copyright," wrote the majority. "Courts should presume that a person who authorizes an activity does so only so far as it is in accordance with the law" (para. 38). Absent evidence of particular knowledge or involvement, a court would not hold that a person who lends a neighbour a chainsaw authorizes its use for murder or destruction of property. The same should hold for copying devices: computers, cameras, or copy machines have significant productive and non-infringing uses, and the law would not presume that making them available to others constitutes authorization to infringe copyright. Libraries, copy shops, and other venues where such technology is available often make this explicit by posting notices that remind users of the existence of copyright law.

Moral Rights

Section 3 describes only the economic interests of copyright owners. But there is a whole other realm of rights that are personal to "the author"—a term that in the Copyright Act refers not only to writers but to artists, composers, and all other creators of works. "Moral rights," as they are called, exist separate and apart from economic rights, and are presented in sections 14 and 28 of the Copyright Act. They can only be held by creators and their heirs—and, as of 2012, by performers[12] and their heirs—but not by corporations or contracting parties. Moral rights can be waived, but they cannot be assigned in the same way as economic rights.

Moral rights are essentially three: the right of *integrity*, the right of *attribution*, and from section 28.2(1)(b), the right of *association*.

The right of integrity is deemed to be infringed only if, per section 28.2(1), the use damages "its author's or performer's honour or reputation." But how do we tell whether a modification to a work has prejudiced the honour or reputation of the author? A couple of Canadian cases have engaged with this question. In *Snow v. Eaton Centre* (1982), the court accepted the artist's subjective opinion that his reputation had been damaged; in *Prise de Parole v. Guérin* (1996), the court would have required some objective evidence to make a finding for moral rights infringement. Painting, sculpture, and engraving aside—they are treated specially in section 28.2(2)—a claim to violation of moral rights will more likely be successful in court if backed up by objective evidence, perhaps in the form of an independent expert witness, linking the change to the artwork with damage to the artist's reputation or honour.

On the right of attribution, we might ask when mentioning the source is "reasonable in the circumstances," a limitation stated in section 14.1(1). This could be interpreted to mean that an artistic practice such as collage, which by convention does not feature attribution, does not infringe the right of attribution. But there is no case law on this question in Canada. We might also ask, What constitutes "association"? The answer is clear, perhaps, when a song is used to advertise a product or promote a political cause. But what if a piece of art is put on the same wall as the logo for a company sponsor? Again, the courts have not spoken on this subject.

Notably, U.S. copyright law does not include a general moral rights provision—apart from some special provisions in the Visual Artists Rights Act.[13] In fact, the United States has been quite hostile to the whole idea of moral rights, which does, after all, sometimes get in the way of maximum exploitation of a market for a work. On this issue, Canada lies somewhere between France (in which moral rights are perpetual) and the United States.

Moral Rights in Works and Performers' Performances in the Copyright Act

14.1(1) The author of a work has, subject to section 28.2, the right to the integrity of the work and, in connection with an act mentioned in section 3, the right, where reasonable in the circumstances, to be associated with the work as its author by name or under a pseudonym and the right to remain anonymous.

17.1(1) . . . a performer of a live aural performance or a performance fixed in a sound recording has, subject to subsection 28.2(1), the right to the integrity of the performance, and . . . the right, if it is reasonable in the circumstances, to be associated with the performance as its performer by name or under a pseudonym and the right to remain anonymous.

28.1 Any act or omission that is contrary to any of the moral rights of the author of a work or of the performer of a performer's performance is, in the absence of the author's or performer's consent, an infringement of those rights.

28.2(1) The author's or performer's right to the integrity of a work or performer's performance is infringed only if the work or the performance is, to the prejudice of its author's or performer's honour or reputation,

   (a) distorted, mutilated or otherwise modified; or

   (b) used in association with a product, service, cause or institution.

(2) In the case of a painting, sculpture or engraving, the prejudice referred to in subsection (1) shall be deemed to have occurred as a result of any distortion, mutilation or other modification of the work.

(3) For the purposes of this section,

   (a) a change in the location of a work, the physical means by which a work is exposed or the physical structure containing a work, or

   (b) steps taken in good faith to restore or preserve the work shall not, by that act alone, constitute a distortion, mutilation or other modification of the work.

— Copyright Act.

Snow v. Eaton Centre (1982)

Toronto's Eaton Centre commissioned artist Michael Snow to create a sculpture of sixty Canada geese in flight. It held all economic copyright in the work. As part of its Christmas decorations for 1982, the management festooned the geese with red ribbons. Snow went to court, seeking an injunction to have the ribbons removed. He argued that the ribbons made his otherwise naturalistic work look ridiculous and constituted a modification that damaged his honour and reputation. The court accepted his argument. In other words, the court deemed that an artist's subjective opinion about the effect of a change to his work on his reputation is sufficient, so long as it is arrived at reasonably. Note, however, that this case was decided before the addition of section 28.2(2) to the Act: today, a sculptor would not have to make a claim about honour and reputation.

Prise de Parole Inc. v. Guérin (1996)

In 1992 the book publisher Guérin published an anthology for schools entitled (ironically, it turns out) *Libre Expression* (Free Expression). The book included an unauthorized extract consisting of about one-third of Doric Germain's young adult novel *La Vengeance de l'orignal* (The Vengeance of the Moose). Sales of the novel plummeted, and Germain's publisher sued for infringement. Germain claimed that the abridgement of the novel—for example, the removal of descriptions of hunting and fishing methods—was also an infringement of his moral rights.

The court found that Guérin had infringed the reproduction right, and awarded damages. But on the issue of moral rights, the court did not find the author's distress at the abridgement of his novel sufficient to prove damage to his reputation and honour. It dismissed the moral rights claim.

5. USERS' RIGHTS

f this book had been written a decade ago, the title of this chapter would not be "Users' Rights." It would probably be something like "Exceptions to Infringement: Fair Dealing and Other Defences." Until recently, fair dealing was not considered more than a fairly long-shot defence to allegations of infringement. After all, in *Michelin v. CAW* (1996), a case concerning the use of an unauthorized image of the Michelin Man on unionization posters, a court ruled that fair dealing provisions "should be restrictively interpreted as exceptions." The court asserted that using what it called "another's private property" to ground one's own original expression was "a prohibited form of expression."

But *Théberge v. Galerie d'Art du Petit Champlain* (2002) marked the beginning of a sea change. In this case, the Supreme Court held that the proper balance in copyright "lies not only in recognizing the creator's rights but in giving due weight to their limited nature. In crassly economic terms it would be as inefficient to overcompensate artists and authors for the right of reproduction as it would be self-defeating to undercompensate them" (para. 31). While the *Théberge* case did not focus directly on fair dealing, the court announced what amounted to a new policy framework for copyright analysis, stating: "Excessive control by holders of copyrights and other forms of intellectual property may unduly limit the ability of the public domain to incorporate and embellish creative innovation in the long-term interests of society as a whole, or create practical obstacles to proper utilization" (para. 32).

Meanwhile, a long-standing dispute between the Law Society of Upper Canada (LSUC) and a group of commercial publishers of law-related materials (including CCH Canadian Ltd.) was working its way through the courts. The Law Society's Great Library had a practice of faxing articles and other research materials to members off-site and of allowing in-house patrons to use self-service photocopiers; the law publishers alleged that these practices constituted copyright infringement. In its 2004 decision in *CCH v. Law Society of Upper Canada*, the Supreme Court declared these uses to be fair dealing. More importantly, the court clarified the nature and character of the fair dealing defence in resonant terms:

> Procedurally, a defendant is required to prove that his or her dealing with a work has been fair; however, the fair dealing exception is perhaps more properly understood as an integral part of the *Copyright Act* than simply a defence. Any act falling within the fair dealing exception will not be an infringement of copyright. The fair dealing exception, like other exceptions in the *Copyright Act*, is a user's right. In order to maintain the proper balance between the rights of a copyright owner and users' interests, it must not be interpreted restrictively. (para. 48)

The court went on to quote, and adopt, legal scholar David Vaver's observation: "User rights are not just loopholes. Both owner rights and user rights should therefore be given the fair and balanced reading that befits remedial legislation."

The same year, in *SOCAN v. CAIP*, the Supreme Court noted, "The capacity of the Internet to disseminate 'works of the arts and intellect' is one of the great innovations of the information age." The court added, "Its use should be facilitated rather than discouraged, but this should not be done unfairly at the expense of those who created the works of arts and intellect in the first place."[1] In these three major cases (*Théberge*, *CCH*, and *SOCAN*), the Supreme Court articulated the idea of users' rights, while also demonstrating the need for careful balancing of interests.

Despite the resounding rhetoric of these cases, or perhaps because of it, fair dealing became quite a battleground of public debate for the ensuing years. But 2012 brought confirmation of the importance of users' rights in the form of both legislation and further jurisprudence. In this year, Parliament added three new accepted purposes for fair dealing—education, parody, and satire—and the Supreme Court reiterated its position that fair dealing is an important users' right in both *Alberta v. Access Copyright* and *SOCAN v. Bell*. There is now no doubt that the term "users' rights" names a weighty counterpart to "owners' rights" and allows us effectively to imagine copyright as a system of relationships and interests. This system does not have to be conceived of as a battle between cultural workers and consumers. We all play various different roles with respect to copyright works, especially in an era in which we are so easily enabled by technology to cut and paste, whether we do so for artistic purposes, for financial gain, for work, for play, for school, or for some combination of reasons. In the post-*CCH* era, and especially now with the additional clarity provided by *Alberta* and *Bell*, Canadians should be just as familiar with the idea of fair dealing as they are with the idea of copyright infringement.

The Statutory Basis of Fair Dealing

Section 29 of the Copyright Act states, "Fair dealing for the purpose of research, private study, education, parody or satire does not infringe copyright." Sections 29.1 and 29.2 also list "criticism or review" and "news reporting" as allowable purposes—with the added requirement of naming the source. The eight categories, then, that fall under the fair dealing provision are research, private study, education, parody, satire, criticism, review, and news reporting. Three of these purposes (education, parody, and satire) were added in 2012.

The list is categorical: that is, before a user can rely on the fair dealing doctrine, that use must fall within one of those eight enumerated categories set out in the Act. This requirement runs counter to popular belief. People often invoke fair dealing in conversation by noting that they have used only a very little of the source, that they have substantially altered it, or that they were not making any money from the use. These factors all come into fair dealing analysis, but they are only relevant if the use first falls within one of the enumerated categories. If the use is substantial (the meaning of which has to do with more than quantity, as we saw in chapter 3), and it falls into one of these categories, then we turn to the *CCH* case for guidance about how to evaluate the fairness of the dealing.

Fair Dealing versus Fair Use

Before we get to the *CCH* fair dealing tests, it is worth noting that much confusion about fair dealing arises from half-familiarity with U.S. "fair use," a similar but distinct legal provision. Section 107 of the U.S. copyright statute states:

> Fair use of a copyrighted work, including such use by reproduction in copies or phonorecords or by any other means specified by [section 106], for purposes *such as* criticism, comment, news reporting, teaching (including multiple copies for classroom use), scholarship, or research, is not an infringement of copyright. [Emphasis added.]

The term "such as" in the U.S. legislation indicates that the uses listed are meant to be illustrative, not exhaustive—whereas the Canadian Act, lacking a "such as" clause, presents the complete list of fair dealing purposes. It is because of the phrase "such as" that case law in the United States has come to recognize parody as a legitimate purpose for fair use, even though it is not named in the Act; in Canada, parody had to be added to the fair dealing list in order to be covered.

The Canadian statute does not define "fair dealing" beyond stating the categories in which it falls, but section 107 of the U.S. statute provides a test for fairness:

> In determining whether the use made of a work in any particular case is a fair use the factors to be considered *shall include—*
> (1) the purpose and character of the use, including whether such use is of a commercial nature or is for nonprofit educational purposes;
> (2) the nature of the copyrighted work;
> (3) the amount and substantiality of the portion used in relation to the copyrighted work as a whole; and
> (4) the effect of the use upon the potential market for or value of the copyrighted work.
> The fact that a work is unpublished shall not itself bar a finding of fair use if such finding is made upon consideration of all the above factors. [Emphasis added.]

Here again the language is open-ended. The four factors are preceded by the important signal words "shall include," leaving courts free to consider other factors they deem pertinent.

That said, prevailing judicial attitudes in the United States have for some time placed a heavy emphasis on the economic interests of copyright owners at the expense of the other fair use factors.[2] And with lawsuits much more common in the United States than they are in Canada (for one thing, there is more money at stake), many scholars and public interest advocates have noted the pronounced chilling effect produced by the mere threat of litigation, whatever the outcome in court might be.[3] Stories of such incidents can mislead Canadians about the risk of practising fair dealing, given our Supreme Court's robust defence of fair dealing in recent years. But perhaps U.S. courts are catching up to our courts: two recent U.S. cases we discuss in chapter 16 seem to have reinvigorated fair use.[4]

The Tests for Fair Dealing

While the Copyright Act itself does not contain further guidance on what does or does not constitute fair dealing in any particular situation, the *CCH* case did. The court said, "The *Copyright Act* does not define what will be 'fair'; whether something is fair is a question of fact and depends on the facts

Many Canadians understand users' rights in terms of the practice of their community or trade, rather than in terms of the law. Sometimes this common sense is in tension with the law, but non-lawyers often apply impressive amounts of self-consciousness and reason to users' rights. Consider the thoughts of art gallery curator Jan Allen:

> I feel in a way, even though we're a public institution, we have a mandate for research and we push those limits. We take certain risks and we work with things that we think are meaningful . . . and part of that is flirting with those edges around the interpretation of copyright. Again, I'm really going by what is common practice and feeling that that is where I have to draw the line. Otherwise . . . if you are too legalistic about it I think you could get off in some weird pristine edge of things and it would be very lonely. Like we've got the painting that came into the collection in the last year that makes reference to Captain Nemo's underwater cave . . . are we going to get permission for that? No, we're not. We're in a place in culture where culture feeds on itself . . . its own codes . . . so you can't start excluding all those codes . . . they are part of the language . . . so I think that you have to trust that the interpretation of the law and that its enactment, its practice, does play itself out so that there's some leeway.

Source: Jan Allen, Curator of Contemporary Art, Agnes Etherington Art Centre, Kingston, Ontario, interview with Kirsty Robertson.

of each case" (para. 52). To analyze these facts, the court laid out a two-step test with six fairness factors that today stands as law.[5]

The first step is the *purpose* of the use. Does it constitute research, private study, education, parody, satire, criticism, review, or news reporting? Neither

Is it fair dealing?

The first step is to consider the purpose of the dealing: Does the use fall under one of the eight categories of research, private study, education, parody, satire, criticism, review, or news reporting? Keep in mind that while the eight categories are to be broadly construed, the use must still fall into one of these categories or it is not fair dealing.

If the use does fall into one of these categories, proceed to the second step, the fairness tests, and make a judgment based on the cumulative weight of all results.

- The purpose of the dealing
- The character of the dealing
- The amount of the dealing
- Alternatives to the dealing
- The nature of the work
- The effect of the dealing on the work

The first fairness test, for the purpose of the dealing, is perhaps redundant with the prior step of categorization. But the Supreme Court seems to be inclined to make the first step quite easy, and then to return to the question of purpose in more detail to confirm its fairness, as we describe in this chapter.

When it comes to the basics of fair dealing, in most situations you will be able to think through the factors and issues yourself, following the elaboration on the six-part test in this section—whether you want to know what uses you can make without permission, or whether you need to know if someone else is infringing on your copyright. If you do have to consult a lawyer or institutional copyright officer, this initial independent analysis will get you off to a good start.

the Act nor *CCH* defines these terms, so it would seem that their definitions would come from ordinary practice and usage. The *CCH* court stated, "These allowable purposes should not be given a restrictive interpretation or this could result in the undue restriction of users' rights" (para. 54).

Once a use has been found to fit one of the categories, the Supreme Court has proceeded to *further explore its purpose* as the first fairness test. In *CCH*, while the court noted that commercial purposes might tend to be less "fair" than non-commercial purposes, it did not disqualify them from this test, stating, "Lawyers carrying on the business of law for profit are conducting research within the meaning of s. 29 of the *Copyright Act*" (para. 51). In both *Alberta v. Access Copyright* and *SOCAN v. Bell*, the court used the purpose test as an occasion to assert that it is the ultimate user's purpose that is primarily at issue here, not an intermediary's purpose. Thus, a student or prospective consumer doing research would fit the purpose test, even if an intermediary had other purposes. Basically, the fairness test for purpose seems to focus on the particular facts of the situation, and to see where the weight of different parties' purposes stands in terms of fairness.

CCH glosses the second test, the *character* of the dealing, as follows:

> If multiple copies of works are being widely distributed, this will tend to be unfair. If, however, a single copy of a work is used for a specific legitimate purpose, then it may be easier to conclude that it was a fair dealing. If the copy of the work is destroyed after it is used for its specific intended purpose, this may also favour a finding of fairness. (para. 55)

These sentences suggest a focus within the character criterion on the number of copies made, and a fairly narrow conception of fair dealing. But the court goes on:

> It may be relevant to consider the custom or practice in a particular trade or industry to determine whether or not the character of the dealing is fair. For example, in *Sillitoe v. McGraw-Hill Book Co.* (U.K.), [1983] F.S.R. 545 (Ch. D.), the importers and distributors of "study notes"

that incorporated large passages from published works attempted to claim that the copies were fair dealings because they were for the purpose of criticism. The court reviewed the ways in which copied works were customarily dealt with in literary criticism textbooks to help it conclude that the study notes were not fair dealings for the purpose of criticism.

This is a key part of the *CCH* decision: it reminds people in diverse trades or industries that their "practice" has a role in determining the law. It would follow that if we narrow our practice, asking permission for every tiniest use, we narrow our users' rights in the law. Instead, we could heed this passage and use past practice, common sense, and a little courage about the habits of use in our particular circles. If, for example, journalists can quote certain portions of printed texts without permission, why can't they do the same with digital texts? If they do not, they are effectively participating in a narrowing of fair dealing.

With the third fairness test laid out in *CCH*, the *amount* of the dealing, the court begins by reminding us of the substantiality requirement for infringement: "If the amount taken from a work is trivial, the fair dealing analysis need not be undertaken at all because the court will have concluded that there was no copyright infringement." The substantiality requirement effectively functions as a users' right. The court observes, "The quantity of the work taken will not be determinative of fairness, but it can help in the determination. It may be possible to deal fairly with a whole work. As Vaver points out, there might be no other way to criticize or review certain types of works such as photographs" (para. 56). This point may surprise many readers: there is no absolute measure or formula for how much or what portion of a work can be used under fair dealing. In this holistic approach to analysis, you simply need to be able to show that you needed to use the amount you used.

The fourth test is *alternatives* to the dealing. Did you really need to use this particular piece for your purpose? The court says, for example, "If there is a non-copyrighted equivalent of the work that could have been used instead of the copyrighted work, this should be considered by the court" (para. 57). On this factor the court also stated that "the availability of a licence is not relevant to deciding whether a dealing has been fair" (para. 70).

Next the court turns to the *nature* of the original work. The examples provided all have to do with publicity and exposure:

> Although certainly not determinative, if a work has not been published, the dealing may be more fair in that its reproduction with acknowledgement could lead to a wider public dissemination of the work—one of the goals of copyright law. If, however, the work in question was confidential, this may tip the scales towards finding that the dealing was unfair. (para. 58)

These comments, in which wide dissemination is seen as a public benefit, sharply contrast with the court's comments on the character of the dealing, in which multiple copies are viewed suspiciously. In any case, this criterion offers a place for an argument about the desirability of further dissemination: users could argue, for example, that reproduction of some portion of an out-of-print book for an enumerated purpose serves the public interest. Conversely, this criterion could provide a basis for an argument that in the case of particular types of works, even small amounts of unauthorized use may be unfair, although this is not a path that the court pursued in *CCH*.[6]

In its discussion of the final fairness test, the *effect* of the dealing on the work, the court is insistent that "although the effect of the dealing on the market of the copyright owner is an important factor, it is neither the only factor nor the most important factor that a court must consider in deciding if the dealing is fair" (para. 59). This will also come as a surprise to most readers: competition in the original's market is only one in the long list of issues. If a use does compete in the market with the original, but is legitimate criticism and only takes what it needs to achieve its purpose, for example, there may well be a finding of fair dealing. Commercial use is not a nail in the fair dealing coffin; the *CCH* case, after all, concerned use by lawyers in private practice who might otherwise have bought more materials from the publishers, and yet the court considered their use fair dealing.

Alberta (Education) v. Access Copyright (2012)

Access Copyright had sought a tariff at the Copyright Board for photocopying of works in its repertoire in elementary and secondary schools throughout Canada outside of Quebec. The main contested issue on appeal was the status of copies that were made at the initiative of teachers to be given to students with instructions to read the materials. The Board agreed with Access Copyright that these copies did not constitute fair dealing and were therefore compensable copies, subject to a royalty. The Federal Court of Appeal upheld the Board's conclusion that the copies were not fair dealing. On review, the Supreme Court of Canada reversed the decision, applying the principles from *CCH v. Law Society of Canada* to the educational setting. Emerging out of this case (and its companion cases) are some very clear general principles:

- The first step and the first fairness test, both concerning the purpose of the dealing, are to be approached with a very broad understanding of the stated categories. The court said that research is not limited to formal and systematic inquiry but equally applies to consumer research, lifelong learning, and everyday information-seeking activities, and that the "private" in private study should not be taken to mean isolation or solitude. The fact that a reading has been assigned by a teacher does not preclude deeming it to be private study.
- The point of view is to be that of the end-user (say, a student) rather than of an intermediary (say, a teacher). While "copiers cannot camouflage their own distinct purpose by purporting to conflate it with the research or study purposes of the ultimate user" (para. 21), a person copying material in order to enable another person's fair dealing with no ulterior motive (such as financial gain) is generally not infringing copyright.
- The test concerning amount of copying is to be understood as "an examination of the proportion between the excerpted copy and the

entire work, not the overall quantity of what is disseminated" (para. 29). In *Alberta*, multiple copies of excerpts from textbooks were considered fair.

- As to the test regarding alternatives, the possibility of simply purchasing the text for everyone (or having everyone purchase a text) when only portions are to be used was not considered reasonable in the educational setting. Purchase policies can be justified where they are reasonably necessary to achieve the ultimate purpose of the students' research and private study.

- On the last test, the effect of the dealing on the work, the court stated that demonstrable harm needs to be shown in order to turn this factor against the fair dealing claimant. Generalizations about lost sales due to copying will not suffice.

Fair Dealing in 2012

While there has been much debate and anguish over the *CCH* case—universities have been overly cautious in embracing its possibilities, and creator groups have inveighed against it—the Supreme Court affirmed it in no uncertain terms in 2012 in both *Alberta v. Access Copyright* and *SOCAN v. Bell*. With the new fair dealing categories added to the Copyright Act and such emphatic Supreme Court cases, we can say that fair dealing is definitely here to stay. Creator and cultural industry groups and their collectives are quite concerned about this development, saying that it will mean lost revenues. But it is not at all clear to us that robust fair dealing will lead to reduced sales of entire works. For example, if teachers rely on fair dealing to provide students with chapters of books or clips from videos, they cannot be said to be avoiding a plausible alternative because they did not assign all the whole works from which each selection is taken. That would be too expensive and not appropriate to many teaching situations. From exposure to selections, students will at least know about those works and perhaps seek them out in the future, much in the way that, as the Supreme Court pointed out in *SOCAN v. Bell*, those who listen to clips on iTunes will buy some of them in their entirety.

However, we do acknowledge that this new world of fair dealing is likely to disturb the existing licensing arrangements administered by collectives. We sincerely hope that collectives rearrange their business models so that they can serve both consumers and their members, an issue that will be discussed further in the next chapter.

Other Exceptions to Infringement

Other sections in the Copyright Act limit the enforceability of owners' rights to various degrees by naming specific acts that do not constitute infringement.[7] Exceptions of general application are listed in Table 8. They include new provisions for non-commercial user-generated content (UGC, discussed further in chapter 10), for making certain personal copies, for fixing broadcasts for later viewing, and for making backup copies. There are also new exceptions to do with backing up computer programs, interoperability, and encryption research. Exceptions for makers of film, video, and photography are reviewed in chapter 11, and those for educational institutions, libraries, museums, and archives are outlined in chapters 15 and 16.

As useful as these specific exceptions may be in particular circumstances, they all need to be approached with caution. They contain counter-exceptions and limitations, so that the users' right described is often quite narrow. For example, while section 29.22 allows you to copy a legally obtained song from one platform or medium to another, it specifically does not cover giving a copy to, say, the other members of your band. But that use could be fair dealing for purposes of criticism or review. It is important to bear in mind that the special exceptions do not displace fair dealing; they are simply supplements that provide the user additional protections in some instances.

Table 8. Exceptions to Owners' Rights of General Application

Common Name	What It Allows	Section of Copyright Act
non-commercial user-generated content	use of an existing work (or other subject matter or copy) in the creation of a new work (or other subject matter); use of the new work, or authorization of an intermediary to disseminate it *but*: use is limited to non-commercial purposes; use cannot have a substantial adverse effect on the (potential) exploitation of the existing work (or other subject matter) or on its existing or potential market	29.21 (new 2012)
reproduction for private purposes	reproduction of a work (or other subject matter) for the copier's private purposes *but*: the source work must be legally obtained, not borrowed or rented, and not an infringing copy; exception inapplicable if TPM circumvented; and new copy cannot be given away	29.22 (new 2012)
reproduction of broadcast for later listening or viewing	recording a broadcast program for the purpose of listening to or viewing it later *but*: not applicable to "on-demand service"; program must be received legally; only one copy can be made and only for private purposes; copy can't be kept longer than reasonably needed for later viewing; exception inapplicable if TPM circumvented; new copy cannot be given away	29.23 (new 2012)
backup copies	reproduction by an owner or licensee of a lawful source solely for backup purposes *but*: exception inapplicable if TPM circumvented; new copy cannot be given away	29.24 (new 2012)
backup of computer program	copying of a computer program by a person who owns it or has a licence to use it, in order to modify/convert it to another computer language if necessary for compatibility or backup	30.6

interoperability of computer programs	copying of a computer program by a person who owns it or has a licence to use it, for the sole purpose of obtaining information that would allow interoperability of that program and another computer program	30.61 (new 2012)
encryption research	copying for the purposes of encryption research if it would not be practical to carry out the research without making the copy and if the person informs the copyright owner	30.62 (new 2012)
security	copying in order to assess the vulnerability of a computer, system, or network or to correct security flaws	30.63 (new 2012)
temporary reproductions for technological processes	reproduction of a work or other subject matter if (a) the reproduction forms an essential part of a technological process; (b) the reproduction's only purpose is to facilitate a use that is not an infringement of copyright; and (c) the reproduction exists only for the duration of the technological process	30.71 (new 2012)
perceptual difficulties	for a person with a perceptual disability (or person acting for them) to reproduce, publicly perform, translate, or adapt a copy or sound recording of a work in a format specially designed for persons with a perceptual disability *but:* not applicable to films; does not authorize the making of a large print book; and does not apply where the work or sound recording is commercially available.	32
lecture reporting	making or publishing, for the purposes of news reporting or news summary, a report of a lecture given in public *but:* not applicable if prohibited by conspicuous written or printed notice	32..2(1)(c)

Table 8. Exceptions to Owners' Rights of General Application (continued)

Common Name	What It Allows	Section of Copyright Act
public reading	reading or reciting in public a reasonable extract from a published work	32.2(1)(d)
political speech reporting	making or publishing, for the purposes of news reporting or news summary, a report of an address of a political nature given at a public meeting	32.2(1)(e)
use of commissioned photograph	for a person who commissioned a portrait or photograph to use it for personal non-commercial purposes *but:* parties can agree the section does not apply	32.2(1)(f) (new 2012)
performance at agricultural fairs	the playing of live or recorded music, or music via radio, without motive of gain at an agricultural or agricultural-industrial exhibition or fair that receives a grant from or is held by its directors under federal, provincial, or municipal authority	32.2(2)
religious, educational, or charitable performance	for a religious organization or institution, educational institution, or charitable or fraternal organization to play live or recorded music, or music via radio, in further-ance of a religious, educational, or charitable object	32.2(3)
private music copying	reproduction of music onto an audio recording medium for the private use of the person who makes the copy	80

6. COLLECTIVES AND THE COPYRIGHT BOARD

P art VII of the Copyright Act provides for the functioning of copyright collective societies, usually known simply as "collectives," regulated by the Copyright Board. Canada has some thirty-four copyright collectives, each representing authors and owners with respect to particular rights in certain media for different types of use.[1] Collectives do not themselves own rights; they act on behalf of rights owners. For example, the Society of Composers, Authors and Music Publishers of Canada (SOCAN) represents performing rights in musical works on behalf of composers and publishers; Audio Ciné and Criterion each represent distinct stables of Canadian and international film studios; CCLI (Christian Copyright Licensing International) serves the licensing needs of Christian publishers; SCAM (Société civile des auteurs multimédia) negotiates broadcast rights on behalf of international francophone creators; and DRTVC (Direct Response Television Collective) licenses the retransmission of infomercials. In many areas of licensing, two collectives exist, one for Quebec and one for the rest of Canada.[2]

Individuals may sometimes interact with collectives in seeking rights clearances. If you want to record your rendition of a song that someone else wrote, you might contact the Canadian Musical Reproduction Rights Agency (CMRRA) and see if your chosen song is in their "repertoire"— that is, if they represent its composer and lyricist. Through reciprocal relationships, Canadian collectives represent rights owners in many countries besides Canada. Collectives do not represent all rights owners—so you might have to pursue your search for the owner by more creative means. But it is easy to search a collective's database to see if it can help with your chosen material.

Such case-by-case or "transactional" licensing is only a part of what collectives offer; the bulk of collectives' transactions are done through blanket licences or tariffs. Many schools, colleges, and universities have maintained licences with Access Copyright, a Canadian agency founded by groups of creators and publishers to license "reprography," or photocopying. These licences have covered many routine copying uses in those institutions according to formulas agreed upon by the parties.

In situations where the parties cannot agree on terms of a licence, which is essentially a private contract, section 70.2(1) provides that "either of them or a representative of either may, after giving notice to the other, apply to the [Copyright] Board to fix the royalties and their related terms and conditions" through a tariff. For example, radio stations and dance halls operate under tariffs with SOCAN for airplay of recordings.[3] In 2006 SOCAN succeeded in having the Board certify a tariff for mobile phone ringtones—which was subsequently appealed and upheld. The Copyright Board functions like a lower court in that it takes evidence on disputed issues of fact and renders decisions. The decisions of the Board are subject to judicial review, first to the Federal Court of Appeal. After that the Supreme Court has the discretion whether or not to grant leave for further review. All five of the 2012 Supreme

Court copyright cases arose out of appeals of Copyright Board cases.

Tariff proceedings are extremely slow, expensive, and technical, and often find themselves overtaken by new developments in technology or law. Creators and users of copyright materials are far away and little heard among all the lawyering. In November 2012, the Copyright Board instituted a review of its procedures and processes to make its operations more efficient and less costly, and generally to improve service to collectives, copyright users, and the general public; we certainly hope these improvements materialize.[4]

When collectives are working well, they make things easier for creators, other rights owners, and users alike. They may be a special help to non-corporate rights holders or rights seekers, who have neither the time nor the expertise to pursue independent searches and negotiations. For example, small publishers don't have rights departments and can benefit hugely from collectives' clearing-house functions. But collectives can be useful for larger institutions and businesses, too. Brian Lamb, Director of Innovation at Thompson Rivers University, observes that it is extremely expensive for individual universities to run their own rights-clearance operations, and if collectives were to offer reasonable terms, avoiding surveillance and overreaching of the law, educational institutions might find good reason to license use through them.[5] Even Cory Doctorow, a science fiction writer well known for his strong users' rights advocacy, acknowledges collectives' utility: "Having a place where you collect some money . . . is not an insane idea . . . overall the efficiencies that you realize in a system like that are greater than the costs. . . . I'm all for that kind of blanket licence solution . . . it will pay artists without criminalizing users."[6] The Songwriters' Association of Canada has been working for some time on a model to monetize file sharing through collectives, proposing a monthly licence fee collected by Internet service providers: "By monetizing behavior rather than any specific technology," they suggest, "music creators and rights-holders will lay the foundations for a business model that can continue for decades rather than attempting the almost impossible task of trying to monetize the ever shortening cycle of changing technology.... A monetized music file-sharing system would give consumers access to the world's entire catalogue of recorded music, and at the same time fairly compensate creators and rights-holders."[7]

Despite such hopes and affirmations, collectives have in practice proven controversial with both users and creators. Payment formulas can put creators at a disadvantage. The music collectives remunerate rights holders based on radio play and album sales, which may mean underpayment to musicians whose work circulates on the Internet or is broadcast on college

In June 2012, Brian Brett, a poet and fiction writer and the former chair of the Writers' Union of Canada, circulated a public letter stating that he "can no longer advocate for Access Copyright. It is time instead for a class action lawsuit by writers against Access Copyright, which has been collecting our money in our names and yet failing to deliver that income equitably and transparently." In the letter, Brett set out a list of eight specific concerns about Access Copyright. He observed, for example, that "after its expenses (which are high—spending approximately $10 million to collect $23.5 million in distributable income), there is $23.5 million in money for copyright distribution. Over $6 million dollars go to foreign copyright organizations. Very little, if anything, is paid back for usage of Canadian copyright material by these organizations." He criticized a lack of transparency in that payments to authors via publishers are not tracked and often, it appears, not delivered. Brett critiqued not just the particular collective, but the legislation behind it: it irks him, for example, that even those authors who may not choose to join Access can collect nothing higher than Access rates for their work. He concluded: "While publishers clearly profit from the current situation, it is actually not their fault. The problem is structural within Access Copyright—its constitution and the make-up of its board, and the anti-creator policies it has chosen to adopt over the years."

Source: Brian Brett, "An Open Letter on Access Copyright and the Canadian Copyright Emergency," *The Tyee* (Vancouver), 28 June 2012.

radio. Access Copyright has long been criticized for unfair distribution practices. Writer D.C. Reid criticized the drop in base payments to writers in 2011: "Last year the baseline was $175. That's all. This does not comprise meaningful income. 80% of writers got less than the previous year's baseline of $612, also a figure that does not comprise meaningful income. The point is that schools, colleges and universities think they are paying money to writers, but they are not. Writers get virtually nothing. Writers don't like this but the reprographic corporation primarily reflects the interests of large educational publishers even though the copying payment was introduced for writers."[8]

One music collective, the Canadian Private Copying Collective (CPCC), has generated special public controversy. The CPCC, as authorized in sections 81 to 86 of the Copyright Act, collects levies on blank recording media on the premise that they are being used for "private copying" of music.

Consumer discontent here has had a number of dimensions. A consumer who pays ninety-nine cents for an iTunes song, and then a levy for the privilege of copying that music onto a CD, is paying twice, and probably doesn't like that. Then, too, although the blank media levy was introduced at the record labels' behest in 1997 as a payment scheme for private copying, the labels have been (unsuccessfully) trying to sue downloaders for doing exactly what the levy is supposed to legitimize.

Consumer resentment has also built up because purchasers of such media must pay a levy to the CPCC for redistribution to music rights owners, even if they are using the media for computer backups or storing non-copyright work. Finally, because of the nature of the music marketplace, most of the money leaves Canada—and other countries do not have equivalent levies that send money back our way.

Sources: www.cpcc.ca; *BMG Canada v. Doe* (2005); Howard Knopf, Testimony to the Canadian Heritage Committee, Ottawa, 20 April 2004.

On the user side, one might note the widespread trend of "upward creep" in tariff levels with collectives: when collectives request a raise in the tariff level, the Copyright Board may give them less than they ask, but will rarely give them nothing—even if there is little justification for an increase. As we explore further in chapter 15 on Education, prices are not the only problem: tariffs may contain other terms that are problematic for users. That said, while pricing for tariffs has reflected a narrow interpretation of fair dealing and other users' rights, the 2012 Supreme Court decisions, all of which arose out of decisions of the Copyright Board, may ensure that the Copyright Board is more attuned to users' rights in future.

7. DETERMINING OWNERSHIP

O ne of the most daunting challenges of working with copyright material is identifying its owner. We have seen that copyright interests are multiple, but that together they (the whole bundle of sticks, as it were) generally do start out in the possession of one owner. It might seem that finding the first owner, at least, would be easy. But in many cases, it isn't. Initial ownership can turn on the terms of employment or contract, the origin of the material, or the nature of the work in question. Once the rights have been transferred to others, things become more difficult still. There is no central registry of copyright. Librarians, copyright collectives, and the Copyright Board can help you in a search for downstream rights holders, but your own research skills are your main tool.

First Ownership

Who is the initial owner of the economic copyright in works set out in section 3? We might presume that it would be "the author"—a term that the Copyright Act uses to refer not only to writers but also to creators of all types of works. Indeed, section 13(1) provides that "the author of a work shall be the first owner of the copyright therein."

Yet in many cases, the author is not the first owner of the economic rights in the work. Section 13(3) states:

> Where the author of a work was in the employment of some other person under a contract of service or apprenticeship and the work was made in the course of his employment by that person, the person by whom the author was employed shall, in the absence of any agreement to the contrary, be the first owner of the copyright.

In terms of copyright as policy, this limitation on the rights of the author is an important part of the Act. Many of the policy justifications for copyright are based on the idea of just deserts for the author-as-copyright-owner, but this section reminds us that many copyright works do not even start out under the ownership of their creator. Memos, promotional material, manuals, company artwork, and the like are all owned by the businesses whose employees generated them. For workers and freelancers, this section raises three practical issues. What does "employment . . . under a contract of service or apprenticeship" mean? How do we determine whether work is done within the "course of employment"? And what does "any agreement to the contrary" mean—can agreements to the contrary be implied, or do they need to be in writing?

As it turns out, the practical applications of section 13 differ in various employment situations (see Part III of this book), but we need to mention a couple of general points here. First, the default rules are just that. In many cases, the parties have decided to order their affairs in a different manner—either through industry practice, or through specific terms of contract. For example, university faculty customarily own the copyright in their work; the custom is often only recognized in writing in situations when it is being overridden.

Second, other areas of law inform some of the legal questions that arise in the context of employment. Questions from fields as diverse as tort liability, taxation, and agency often turn on whether or not someone was employed and whether they were acting in the scope of their employment when a particular incident or event occurred. These precedents can be applicable to the issues raised under section 13. For example, freelancers are not considered to be employees even if they are working under a contract.

Even when the economic rights pass to the employer under section 13(3), the first owner of *moral* rights is always the author. However, since many employers require that moral rights be waived, moral rights often do not exist in work produced in the course of employment.

One final possibility for first ownership is the Crown. Under section 12 of the Copyright Act, where any work has been prepared or published by or under the direction or control of Her Majesty (or any government

Table 9. Rules Governing First Ownership

Subject Matter	Owner's Right	First Owner
works	section 3	general rule: author, section 13(1)
		exception for works made in scope of employment: employer, absent agreement to contrary, section 13(3)
performer's performance	sections 15 and 26	performer, section 24(a)
sound recording	section 18	maker, section 24(b)
communication signal	section 21	broadcaster, section 24(c)

department), the copyright in the work belongs to Her Majesty. This ownership rule is a default, and can be altered by contract (freelancers doing research for the government may sometimes retain copyright to their work, for example). Where copyright is owned by the Crown, its duration is fifty years following the end of the calendar year of its first publication.

In recent years the potential severity of Crown copyright has been lessened somewhat by grants of public licences to use certain types of government documents such as laws, regulations, and court cases.[1] But Crown copyright is increasingly criticized on the grounds that taxpayers who have already paid for work produced by the government ought not to face any costs or hurdles to use it. In the United States, copyright does not subsist in federal government works.[2]

Preparing a Canadian edition of a guide to writing style in 2004, Professor Ira Nadel at the University of British Columbia wanted to include Pierre Trudeau's 1970 address to the nation announcing the War Measures Act and imposing martial law to contain the FLQ crisis. But who owned the rights? His first thought was that a public speech by a prime minister would be public domain. When a colleague told him about Crown copyright, he diligently contacted—in turn—the Prime Minister's Office, the Library of Parliament, the National Archive, and Trudeau's law firm. No one at any of these places knew how he could go about clearing copyright. Finally, through the Trudeau Foundation, and more or less by accident, he heard from an executor of Trudeau's will, who gave him permission to publish.

The irony is that, according to Elizabeth Judge, Associate Professor of Law at the University of Ottawa and an expert on Crown copyright, such a speech is almost undoubtedly owned by the Crown, not the Trudeau estate. Had the speech been given in a less clearly formal capacity, it might have been Trudeau's copyright. But the text of a prime ministerial speech such as this would most likely belong to the Crown.

Assigned Rights

For creators the various ways in which first owners can redeploy their bundle of rights are bread-and-butter information. But these points are equally important for users of copyright material—who are also often creators, too. There is, alas, no central registry where you can find the owner of copyright in a given work you want to use or license: mandatory registration of copyright is not permitted in countries that are members of the Berne Convention. Finding copyright owners can require considerable detective work, but it helps to start with knowledge of the ways in which the rights—the sticks in the bundle—can be dispersed.

Although moral rights, understood to be an extension of the persona of the creator, cannot be assigned, they can be passed on to heirs (for the duration of fifty years after the death of the author), and they can also be waived. In identifying the layers of rights in any work, it is important to know if moral rights exist, and to keep analysis of them distinct from analysis of economic rights. Unfortunately, the only way of being sure if they exist or not is to see the author's or employee's contract.

Economic or section 3 rights, though, are freely assignable because the law favours a fluid and flexible system of multiple transactions with respect to even the same work. Section 13(4) of the Act provides:

> The owner of the copyright in any work may assign the right, either wholly or partially, and either generally or subject to limitations relating to territory, medium or sector of the market or other limitations relating to the scope of the assignment, and either for the whole term of the copyright or for any other part thereof, and may grant any interest in the right by licence, but no assignment or grant is valid unless it is in writing signed by the owner of the right in respect of which the assignment or grant is made, or by the owner's duly authorized agent.

Again, interests may be assigned in whole—that is, the entire interest can be transferred to another party—or in part. A partial assignment may apply to a specific territory, to a particular period of time, to a particular medium, or to a particular aspect of the interest. For example, just because a textbook publisher or a website has permission to reproduce a work does not mean that it can authorize further uses of that work.

As we've seen, performers' performances, sound recordings, and broadcast signals are also covered by copyright law. To determine the copyright in, say, a DVD of a televised concert, it is helpful to think of the metaphor of an onion. Let's start near the centre of the onion. When a performer performs a work, we are dealing with two distinct copyrightable objects: the performer's performance and the work. The performer has economic rights in the performance, as set out in section 15 of the Act. If the work (a song, for example) is within copyright, we have to analyze its rights ownership situation apart from that of the performance—and there may be both economic and moral rights to consider. Now let's add two more layers, as we consider the recording of the broadcast of the performer's performance of the work. We now have two more owners: the maker of the recording and the broadcaster. Each participant in the chain requires permission of those holding rights in prior links, and some uses of the DVD might require separate fees or permissions at each level.

Four final comments in the light of this somewhat daunting situation. First, don't forget that copyright term is limited, in most cases, to fifty years after the death of the author, as we explained in chapter 3. So don't go tearing your hair out for public domain works. Second, with the substantiality requirement as described in chapter 3 and the powerful provision for fair dealing and other exceptions in the Copyright Act discussed in chapter 5, don't assume that you have to get permission for every little molecule of copyright material you use. Third, don't assume that any request for permission will be met with a demand that you remortgage your house. If your proposed use is only modestly remunerative, or it's really interesting, the owner may be willing to negotiate a reasonable price or even no price at all. This is, of course, less likely with Disney than with the painter down the road, but acknowledging the wide range of interests among copyright owners can humanize the rights clearance process. And finally, in facilitating rights clearance, you may find some help from one of Canada's many copyright collectives, described in the previous chapter.

Unlocatable Copyright Owners

Let's say that you want to use a work, and you determine that the work is still within its copyright term, that your use is not covered by an exception or otherwise permitted, and that no collective can help you—and you still can't find the copyright owner. Many types of works are not clearly marked with their author's name. Other works may be clearly labelled, but it is impossible to find the author or the author's heirs. Alternatively, the copyright in a given work may have been resold several times and the path may be too faint or too convoluted to trace. If you have searched libraries, bibliographies, phone directories, and other sources and have not been able to find the copyright owner of a work that you want to reproduce or adapt in some way, what do you do?

In Canada, the Copyright Board manages a clearance system for published works whose copyright owners are unlocatable. In order to qualify for such a licence, a user must have made a "reasonable effort" to locate the rights owner.[3] In some cases royalties may be assessed, usually to be paid through the pertinent collective society.

The United States and Europe have no such system in place, and have been studying how to approach the barrier to access and research posed by "orphan works."[4] Some critics consider Canada's system cumbersome or unfair. Why, they ask, should one pay royalties to others when a rights holder cannot be found? And how do we deal with orphaned unpublished works, which lie outside Canada's system and yet are often of the greatest interest? Some have suggested that a better way of dealing with orphan works might be to abolish the licensing system and establish some sort of limit on damages if the follow-on user had looked diligently but unsuccessfully for the rights holder, and the rights holder came forward after the fact. The problem here would be in determining what constituted due diligence. With some media particularly, such as photographs, locating owners is so hard that due diligence is a heavy burden, especially if one is working with a whole collection.

But perhaps the problem is a little simpler now: at least some orphan works reuse can rely on fair dealing. The goal of digitization of library collections is to facilitate research and education—fair dealing purposes both—and digital archives with specific non-commercial terms of use would likely not run afoul of any of the fair dealing tests other than amount, which the *CCH* court said was not in itself decisive (see chapters 5 and 16). We would observe, too, that the problem is not as large in Canada as it is in other jurisdictions, because with a relatively short copyright term of fifty years after the author's death, and the relative simplicity of calculating copyright terms, we have a larger and more transparent public domain.

8. ENFORCEMENT OF
OWNERS' RIGHTS

Copyright cases can be brought on either civil or criminal grounds. Civil infringement actions are lawsuits brought by one private party alleging that another party has engaged in infringing activities with respect to their copyright; criminal cases are prosecuted by the government (known as the Crown). Civil copyright cases are much more common. Both kinds of cases can move up through the courts if they are appealed and cross-appealed. (See Table 10.) In this chapter we explain primary and secondary civil infringement and distinguish them from the Copyright Act's treatment of criminal infringement; we end with a discussion of the 2012 digital locks provisions of the Copyright Act, which are essentially a new enforcement mechanism.

In general terms, the law recognizes two sorts of civil infringement: *general (or primary) infringement* and *secondary infringement.* Section 27(1) of the Copyright Act sets out the definition of "primary infringement": "It is an infringement of copyright for any person to do, without the consent of the owner of the copyright, anything that by this Act only the owner of the copyright has the right to do."

Accordingly, if an aggrieved copyright holder files suit, and if one of the basic owners' rights is implicated, a judge has to decide on the basis of the

Table 10. Civil and Criminal Law Cases

	Civil Law	Criminal Law
parties	plaintiff versus defendant, e.g., *Jones v. Kahn, CCH v. Law Society of Upper Canada*	the Crown versus a private party, e.g., *R. (Regina) v. Jones*
case initiated by	plaintiff	the Crown
source of law	various in different areas of liability; can be based on common-law principles or specific statute	specific statutory offences are enumerated in the Criminal Code and other Acts
evidentiary burden	plaintiff generally has burden of proof on all elements of the claim (e.g., infringement); but burden is usually only at the level of preponderance of evidence; onus then shifts to defendant to establish defence (e.g., fair dealing)	Crown must bear heavier burden of proof (beyond reasonable doubt); in the case of criminal copyright liability, Crown must also negate fair dealing
penalties in Copyright Act	sections 34 through 41	sections 42 and 43
legal costs borne by	plaintiff or defendant	Crown

facts whether or not the owner of the copyright consented. Consent can be implied as well as expressed explicitly, so there might be additional factual issues based on all of the circumstances. Then there is the issue of whether or not the user can defend the infringement—given that many technical instances of infringement do not amount to actionable infringement because of the availability of a defence such as fair dealing. Or perhaps the individual defendants are able to show that they did not have access to the work that they allegedly infringed and did not therefore actually "copy" the work. (It is possible for two people to create identical or highly similar works independently.) Various types of defence can be mounted depending on the specifics of the case. Assuming that there is actionable infringement, the issue then becomes a matter of what remedies are available to the aggrieved owner.

Turning to secondary infringement, section 27(2) of the Copyright Act lays out a set of five actions that fall into this category:

It is an infringement of copyright for any person to
(*a*) sell or rent out,
(*b*) distribute to such an extent as to affect prejudicially the owner of the copyright,
(*c*) by way of trade distribute, expose or offer for sale or rental, or exhibit in public,
(*d*) possess for the purpose of doing anything referred to in paragraphs (*a*) to (*c*), or
(*e*) import into Canada for the purpose of doing anything referred to in paragraphs (*a*) to (*c*),
a copy of a work, sound recording or fixation of a performer's performance or of a communication signal that the person knows or should have known infringes copyright or would infringe copyright if it had been made in Canada by the person who made it.

In other words, secondary infringement is the distribution of infringing works. Significantly, most of the listed circumstances in this section deal with

Planning Ahead; or, Shutting the Barn Door before the Horse Is Gone

The most important way to protect your copyright is to make a contract, or read the contract you are offered, before the work is done. Many people who feel that their work has been stolen or unfairly used find out that they actually agreed to those uses beforehand, or that the law permits them in the absence of a contract.

Contracts do not have to be in legalese to be binding. Even something written on the back of an envelope is better than nothing. Write down what uses of your work you are permitting for the price offered. If you specify particular uses (say, use on a CD cover), the contractor will have to come back to you for any further permissions, such as use of your work on T-shirts, posters, or websites. Sample contract language and terms for artists, designers, and writers can often be found through professional organizations (for example, see writers.ca, a website presented by the Professional Writers Association of Canada, and the Canadian Alliance of Dance Artists' Professional Standards for Dance at http://cadaontario.camp8.org).

If you are handed a long, complicated contract and don't know what it means, don't sign until it's been explained in plain English. Sometimes it is possible to cross things out and adjust them. Other times, of course, it's a take-it-or-leave-it boilerplate. At the very least, be sure to read what you're signing.

issues that arise in the course of trade: owners of copyright would most likely invoke this section as part of an effort to reduce the commercial circulation of pirated materials. The exception is paragraph (b), which does not require the infringer to be engaged in any sort of trade or business. That particular paragraph may have an impact on a wider range of unauthorized uses.

Another important point is that secondary infringement requires an element of knowledge: that is, alleged infringers will not be found liable unless they know (or should have known) that their actions would constitute infringement.¹ One would probably not be held liable for unintentional secondary infringement.

Note that in the case of primary infringement, under section 27(1) there is no such knowledge requirement: liability attaches for general infringement on what is called a "strict liability" basis. In other words, ignorance is no excuse—although as we will see, it can reduce damages to zero.

Assuming that a plaintiff can make out a successful case for infringement, what is that person entitled to collect from the defendant? According to section 34(1), "Where copyright has been infringed, the owner of the copyright is, subject to this Act, entitled to all remedies by way of injunction, damages, accounts, delivery up and otherwise that are or may be conferred by law for the infringement of a right."

An injunction is a court order to cease and desist from the infringing activity. (Under section 39.1, an injunction can be granted even before the case is heard if the court decides that the case is strong enough.) "Delivery up" means forfeiting the offending materials, as detailed in sections 38(1) and 38(2). As for damages, section 35(1) clarifies their possible extent:

> Where a person infringes copyright, the person is liable to pay such damages to the owner of the copyright as the owner has suffered due to the infringement and, in addition to those damages, such part of the profits that the infringer has made from the infringement and that were not taken into account in calculating the damages as the court considers just.

In determining the extent of profits, the plaintiff is required to prove only receipts or revenues derived from the infringement—section 35(2)(a)—and the defendant is required to prove every element of cost that is being claimed—section 35(2)(b). That is to say, profits are determined or estimated with regard to the infringing activity, not to the whole enterprise of either party.

In many situations, then, there would not be much risk of serious liability, because the damage to the owner is minimal and the infringer has gained little

I think my copyright is being infringed. What should I do?

You could

- register your copyright if you haven't already
- send a polite, factual, yet firm letter to the alleged infringer asking for payment or for the material to be withdrawn
- publicize the situation, to shame the infringer or solicit community advice
- consider small claims court
- seek legal counsel if you feel it is warranted

Consider what your goals are: Do you just want the material gone? Do you want a cut of the profits? Is it your moral rights that you feel have been violated? Your strategy will depend on your specific goals and on the nature of the alleged infringer. If it's an individual or a small business, an email in their direction may do the trick. If it's a corporation, you'll probably want a more formal letter with some legalese.

I just got a letter from a lawyer stating that I am infringing his client's copyright. What should I do?

You could

- withdraw the offending material or otherwise comply with the demand
- carefully compose a response letter asking for further clarification and support for the contention that there has been infringement
- carefully compose a response letter explaining why you believe you are not infringing, why your actions constitute fair dealing or some other defence, or why the fees requested are unreasonable
- publicize the situation, to shame the harasser or solicit community advice
- turn it over to legal counsel if you feel it is warranted

Remember, a threatening letter is not a lawsuit. Many rights holders send out hundreds or thousands of such letters, but have no intention of going to court, and thus wield no serious threat to enforce whatever retroactive fees or licences they may demand. See the Canadian Internet Policy and Public Interest Clinic's website at www.cippic.ca/Trolls for information about how to respond to what can often amount to extortion by intermediaries of rights holders.

or no profit. A reproduction of a photograph on a website, for example, would probably not trigger any liability from section 35, because the financial injury to the rights holder and the financial benefit to the infringer would be equally difficult to demonstrate. But section 38.1(1) provides that the copyright owner can elect to recover "statutory damages" instead of the actual damages set out in section 35. If a court awards statutory damages, it has the discretion to set them anywhere between $500 and $20,000 for all infringements at issue in the case of commercial infringements, or between $100 and $5,000 for all infringements at issue in the case of non-commercial infringements, as the court considers just.

The distinction between damage limits for non-commercial and commercial infringements was established in 2012. In debate over the bill in Parliament, Industry Minister Christian Paradis observed:

> Currently, those who have been found to violate copyright can be found liable for damages from $500 to $20,000 per work. If people illegally download five songs, for example, they could theoretically be liable for $100,000. In our view, such penalties are way out of line. As such, the bill proposes to reduce the penalties for non-commercial infringement.[2]

Note especially that the $5,000 cap, set out in section 38.1(1)(b), is for "all infringements involved in the proceedings." Furthermore, section

38.1(5) encourages judges to consider, as they decide the level of damages to be awarded for non-commercial infringement,

(a) the good faith or bad faith of the defendant;
(b) the conduct of the parties before and during the proceedings;
(c) the need to deter other infringements of the copyright in question; and
(d) in the case of infringements for non-commercial purposes, the need for an award to be proportionate to the infringements, in consideration of the hardship the award may cause to the defendant, whether the infringement was for private purposes or not, and the impact of the infringements on the plaintiff.

Section 38.1(2) provides that "if . . . the defendant satisfies the court that the defendant was not aware and had no reasonable grounds to believe that the defendant had infringed copyright, the court may reduce the amount of the award . . . to less than $500, but not less than $200."[3] If the plaintiff is seeking actual as opposed to statutory damages, section 39(1) provides that "if the defendant proves that, at the date of the infringement, the defendant was not aware and had no reasonable ground for suspecting that copyright subsisted in the work or other subject-matter in question," no damages can be awarded. Section 39(2) states, however, that this opportunity for the defendant to avoid damages is not applicable if "at the date of the infringement, the copyright was duly registered under this Act." And there is no guarantee a judge will be easily convinced by a defendant's claim of blissful ignorance.

Often, copyright holders threaten alleged infringers with the possibility of legal action in the hope that a letter alone will produce withdrawal of the offending material. This can be a good strategy for an independent rights holder who doesn't have the time or money to hire a lawyer. But when such a letter comes from a corporate rights holder on law firm letterhead, it can intimidate individuals or small organizations, whether

In 2005, the City of Penticton, British Columbia, chose Michael Hermesh's sculpture *The Baggage Handler* as one of a series of installations in a major traffic circle. Consisting of a somewhat melancholic two-metre-high man carrying—and surrounded by—suitcases, the sculpture provoked some complaints. For his part, the city's mayor didn't realize that its central figure, who came to be known as "Frank," was to be nude, and he ordered the statue removed. After considerable brouhaha in the local and international press, city council voted to keep the statue in place despite its immodesty, but vandals attacked the statue repeatedly in the following weeks. Arguing that the city had been negligent in protecting the statue, Hermesh went to small claims court to recoup the costs of having the sculpture repaired and to seek damages. He was granted $14,386.

Sources: "B.C. Cover-Up," *National Post*, 1 May 2012;
Michael and Carol Hermesh, personal communication.

or not it actually stands for a plausible charge of infringement. Such letters often represent the law of the jungle more than the law of the land. Canada has recently seen a spate of threats triggered by Internet trolls, some of them requesting completely unreasonable fees for the use of individual photographs.[4] In cases in which the alleged infringers have a plausible defence against infringement, in which a lawsuit would produce more bad PR than it would be worth for the corporate rights holder, or in which the use is non-commercial and thus the damages available are insignificant in terms of the corporate bottom line, it often pays to stand firm against such letters.

What standing firm means, though, is the difficult question. Lawyers are extremely expensive. So for any little guy, whether user or creator, using copyright law is a daunting prospect. For example, Catherine Leuthold sued the CBC for 18 seconds' worth of display of her photographs in a documentary about 9/11, and she was awarded $19,200. But this was a tiny fraction of the damages she had sought, and because of her repeated refusal to settle out of court (for an amount that turned out to be higher than what she got in the end), she was required to pay double the CBC's costs for a prolonged trial—and that on top of her own legal costs. Given that she declared her annual income in 2006 to be $20,661, it is unlikely that she can consider her award a victory.[5] Going to court is not for the faint of heart.

One low-overhead option for those wishing modest recompense for a case of infringement is small claims court. The claims limit in such a court ranges from $7,000 (Quebec) to $30,000 (New Brunswick), so this is not a place to win the lottery, but it can meet the needs of many routine disputes. In small claims court, you do not need a lawyer.

Another option if faced with infringement claims or apparent infringement of your own work is to see if your professional organization offers any legal resources or information. Finally, there is the legal clinic route. The box on the next page lists clinics serving artists specifically, and a clinic that may be able to assist those sued for infringement. Pro bono (volunteer) work is not required by Canadian Bar Associations, as it is in the United States, which is one of many reasons such clinics are few and offer quite few services. Most only offer consultation rather than legal representation. But still, it's a start—and sometimes enough.

Criminal Infringement

Section 42(1) of the Copyright Act provides that anyone who knowingly does any of the following is guilty of a criminal offence:

(a) makes for sale or rental an infringing copy of a work or other subject-matter in which copyright subsists,

(b) sells or rents out, or by way of trade exposes or offers for sale or rental, an infringing copy of a work or other subject-matter in which copyright subsists,

Art Law Clinics

Artists' Legal Advice Services (ALAS), Toronto

www.alasontario.com

416-367-ALAS (2527)

Provides 30 minutes of summary legal advice for $20 for artists resident in Ontario

Artists' Legal Outreach, Vancouver

http://artistslegaloutreach.ca

Provides summary legal advice by donation

Artists' Legal Services Ottawa (ALSO)

www.artslawottawa.ca

Provides workshops and consultations

Visual Artists' Legal Clinic of Ontario (VALCO)

www.carfacontario.ca

Provides one hour of free summary legal advice to members of CARFAC (Canadian Artists' Representation) Ontario

Clinique juridique des artistes de Montréal (CJAM)

www.cjam.info

Provides fact sheets and info nights

Artist' Legal Information Society (ALIS), Halifax

http://nsalis.com

Provides consultations, presentations, and a legal information database

Clinic for Users' Rights Issues

CIPPIC (Samuelson-Glushko Canadian Internet Policy and Public Interest Clinic)

www.cippic.ca

Takes on selected cases in the areas of copyright, technology, consumer rights, and public interest

(c) distributes infringing copies of a work or other subject-matter in which copyright subsists, either for the purpose of trade or to such an extent as to affect prejudicially the owner of the copyright,

(d) by way of trade exhibits in public an infringing copy of a work or other subject-matter in which copyright subsists, or

(e) imports for sale or rental into Canada any infringing copy of a work or other subject-matter in which copyright subsists.

Notice the similarity between criminal infringement and secondary infringement in terms of the emphasis on commercial practices. Also like secondary infringement, criminal infringement has a specific knowledge requirement. For criminal infringement, there has to be actual knowledge demonstrated, but for secondary civil infringement the knowledge requirement can be satisfied if defendants "should have known" they were infringing.

Section 42(1) provides the penalties for these offences: if convicted, the offender will be subject "(f) on summary conviction, to a fine not exceeding twenty-five thousand dollars or to imprisonment for a term not exceeding six months or to both, or (g) on conviction on indictment, to a fine not exceeding one million dollars or to imprisonment for a term not exceeding five years or to both." The Crown prosecutor decides, based on the seriousness of the offence, whether to proceed summarily or by indictment. In the world of filmmaking and distribution, Hollywood has argued that Canadian law is inadequate as a means of preventing piracy, but these penalties do seem to be fairly substantial. In addition, section 43(1) defines the criminal offence of commercial public performance of musical and dramatic works, and lays out the penalties for such an action.

Canada, along with many other countries, has come under criticism from the Office of the United States Trade Representative for a variety of copyright-related issues, including criminal enforcement. The actual amount of commercial piracy being committed or facilitated in Canada is the subject of hot debate. In principle, we see no contradiction between taking a generous view of users' rights and enforcing commercial infringement:

it is absolutely crucial to distinguish the legitimate exercise of users' rights from industrial-scale criminal copyright infringement—as far too little public discussion does. At the same time, the criminal enforcement of copyright represents a public subsidy of private business interests—unlike civil litigation, it is paid for by taxpayers. Police, prosecutors, and courts have a raft of immediate public safety concerns that they need to focus on or address and, clearly, decisions about Canada's priorities in the area of copyright protection need to be made in a broad policy context.

Digital Locks

Bill C-11 added highly controversial digital locks provisions to Part IV of the Act. These new provisions are so broad that they could easily be confused for new exclusive owners' rights or new subject matter, but it is most technically correct to characterize them as enforcement measures.

While we use the popular term "digital locks" to describe these provisions, the Act speaks in terms of "circumvention" and "technological protection measures (TPMs)," both of which it defines in section 41:

> "circumvent" means,
>
> > (a) in respect of a technological protection measure within the meaning of paragraph (a) of the definition "technological protection measure", to descramble a scrambled work or decrypt an encrypted work or to otherwise avoid, bypass, remove, deactivate or impair the technological protection measure, unless it is done with the authority of the copyright owner; and
> >
> > (b) in respect of a technological protection measure within the meaning of paragraph (b) of the definition "technological protection measure", to avoid, bypass, remove, deactivate or impair the technological protection measure.
>
> "technological protection measure" means any effective technology, device or component that, in the ordinary course of its operation,

(*a*) controls access to a work, to a performer's performance fixed in a sound recording or to a sound recording and whose use is authorized by the copyright owner; or

(*b*) restricts the doing—with respect to a work, to a performer's performance fixed in a sound recording or to a sound recording—of any act referred to in section 3, 15 or 18 and any act for which remuneration is payable under section 19.

These definitions set the parameters of the prohibition that follows in the next section. A TPM of the type (a) is often referred to as an "access control," because its purpose is to control the access to a work (or other subject matter). The typical access control would involve passwords and other similar restrictions that restrict the ability of the public to gain entry. A TPM of the type described in part (b) is also referred to as a "copy control" or a "use control," because it limits what a user can do once access into a system has been gained. Typical examples include measures that disable features like copying and pasting, printing, saving, and the like.

Given these definitions, the operative provision is section 41.1, which prohibits several acts regardless of whether or not they otherwise constitute copyright infringements. With respect to access controls, section 41.1(1) provides: "No person shall (*a*) circumvent a technological protection measure within the meaning of paragraph (*a*) of the definition 'technological protection measure' in section 41." In addition to prohibiting direct acts of circumvention with respect to access controls, the section 41.1 prohibition goes on to also proscribe indirect acts of circumvention with respect to both copy and access controls:

(1) No person shall ...
(*b*) offer services to the public or provide services if
  (i) the services are offered or provided primarily for the purposes of circumventing a technological protection measure,
  (ii) the uses or purposes of those services are not commercially significant other than when they are offered or provided for the pur-

poses of circumventing a technological protection measure, or

(iii) the person markets those services as being for the purposes of circumventing a technological protection measure or acts in concert with another person in order to market those services as being for those purposes; or

(c) manufacture, import, distribute, offer for sale or rental or provide—including by selling or renting—any technology, device or component if

(i) the technology, device or component is designed or produced primarily for the purposes of circumventing a technological protection measure,

(ii) the uses or purposes of the technology, device or component are not commercially significant other than when it is used for the purposes of circumventing a technological protection measure, or

(iii) the person markets the technology, device or component as being for the purposes of circumventing a technological protection measure or acts in concert with another person in order to market the technology, device or component as being for those purposes.

These provisions are also known as the "service prohibitions" and "device prohibitions," respectively. Regardless of whether the provisions of a service or device in aid of circumvention result in an infringement of copyright, sections 41.1(2) through (4) give the owner of the work or other subject matter a broad range of remedies. In more ordinary parlance, you do not have to be infringing to be banned from getting around digital locks.

Sections 41.11 through 41.18 do go on to provide several exceptions to the prohibitions—for law enforcement and national security, interoperability, encryption research, personal information, security, persons with perceptual difficulties, broadcast logistics, and radio apparatuses. But these exceptions are very narrow and contain numerous counter-limitations and conditions. For example, in the encryption exception, the notice requirement to the owner is unreasonable and unnecessary, and could well impede useful research. The privacy exception is inadequate because of its vague requirement in section 41.14(2) that the device or services "do not unduly impair" the TPM. The same problem limits the effectiveness of the exception for persons with perceptual difficulties. Most importantly, though, there needs to be a general exception for non-infringing uses, as well as for personal uses and for general backup and archiving.

In addition to the anti-circumvention rules, section 41.22(1) provides restrictions with respect to "rights management information (RMI)," which in section 41.22(4) is defined as information that

> (a) is attached to or embodied in a copy of a work, a performer's performance fixed in a sound recording or a sound recording, or appears in connection with its communication to the public by telecommunication; and
>
> (b) identifies or permits the identification of the work or its author, the performance or its performer, the sound recording or its maker or the holder of any rights in the work, the performance or the sound recording, or concerns the terms or conditions of the work's, performance's or sound recording's use.

Removing or altering RMI is prohibited by section 41.22(1):

> No person shall knowingly remove or alter any rights management information in electronic form without the consent of the owner of the copyright in the work, the performer's performance or the sound recording, if the person knows or should have known that the removal or alteration will facilitate or conceal any infringement of the owner's copyright or adversely affect the owner's right to remuneration under section 19.

Note that unlike the prohibitions against circumventing TPMs, this section contains a knowledge requirement.

All in all, we find these new sections of the Copyright Act troubling. While Canada did hold out for many years against the model of the United

States' highly problematic Digital Millennium Copyright Act (DMCA), it ultimately succumbed to the relentless pressures from the U.S. government and copyright industries. We can only hope that, over the coming years, these provisions will be applied and interpreted reasonably. The presence of wording within the new users' rights exceptions that explicitly prohibits the circumvention of TPMs (see Table 8 on pages 84–86) may offer some hope that fair dealing, which does not feature any such wording, may be seen by the courts to prevail over the new digital locks rules.

PART III

PRACTICE

9. MUSIC

Music has been the focus of most of the media hysteria and consumer discontent about copyright. It's a battle of studies and statistics: one week the recording industry seems to have facts on its side indicating that the world of recorded music as we know it is ending, and the next week it doesn't.[1] The recording industry demands protection from the Internet, consumers demand protection from the music industry, and musicians—well, perhaps they don't have much hope that anyone will look out for them. Setting aside the question of who's right and who's wrong, there is clearly a massive disconnect, when it comes to music, between sociocultural norms and emerging artistic forms on the one hand, and the law (or rather what people think the law ought to be) on the other. Music is a hotbed of the copyright legitimacy crisis.

We cannot attempt to cover all of the ins and outs of the situation in one chapter, but we will try to shed light on the extraordinarily complex system of rights ownership and management that emerged in the music industry and copyright law over the course of the twentieth century—a system that is being challenged by the rise of digital technologies.

Twentieth-Century Music Rights

Imagine you are holding a CD in your hand. (If this is an unfamiliar situation, bear with us: it helps to think in material terms to explain the different kinds of ownership of music.) You bought it; you own it. You can play it for your friends, give it away, resell it, copy it for personal use, or hang it on a tree to scare crows. But while you own the tangible thing and have some user's rights in the music embedded in it, other people and companies hold rights in other aspects of the CD: the intangible underlying musical composition, the performance of that music, the sound recording itself, and the cover art and liner notes are all materials in which copyright may subsist.

Let's consider the musical work first.[2] The original composer may own the rights embodied in the particular sequence of notes, rhythms, and chords. A lyricist or an arranger might also have been involved in the creation of the original work. But these creators of the underlying music may have assigned their rights to a publisher or record company; if so, the publisher now owns the creators' rights, for which royalties will be received. The publisher may have required a waiver of moral rights from the creators.

Then there's the performer or performers. Rights in the performer's performance exist separate and apart from the rights in the original work. The performer may have signed with a label—a company that promotes and manufactures records—so it's not always easy to know who exactly holds these rights.

Another neighbouring right rests in the sound recording (or "phonogram"). The "maker" of the master recording—the person who arranged for it to be done—owns the copyright in that master.[3] For commercial music, this maker would usually be the record label. If the CD is a remastered reissue (an old jazz recording digitally cleaned up and enhanced, for

Section 2 of the Copyright Act defines a "performer's performance" as

> any of the following when done by a performer:
>
> (a) a performance of an artistic work, dramatic work or musical work, whether or not the work was previously fixed in any material form, and whether or not the work's term of copyright protection under this Act has expired,
>
> (b) a recitation or reading of a literary work, whether or not the work's term of copyright protection under this Act has expired, or
>
> (c) an improvisation of a dramatic work, musical work or literary work, whether or not the improvised work is based on a pre-existing work

"Performance" is in turn defined as "any acoustic or visual representation of a work, performer's performance, sound recording or communication signal, including a representation made by means of any mechanical instrument, radio receiving set or television receiving set."

Section 15(1) goes on to set out the exclusive rights of the owner of the "performer's performance" (just as section 3 sets out the exclusive rights held by the owner of a "work"):

> . . . a performer has a copyright in the performer's performance, consisting of the sole right to do the following in relation to the performer's performance or any substantial part thereof:
>
> (a) if it is not fixed,
>
> (i) to communicate it to the public by telecommunication,
>
> (ii) to perform it in public, where it is communicated to the public by telecommunication otherwise than by communication signal, and
>
> (iii) to fix it in any material form,
>
> (b) if it is fixed,

> (i) to reproduce any fixation that was made without the perform-
> er's authorization,
>
> (ii) where the performer authorized a fixation, to reproduce any
> reproduction of that fixation, if the reproduction being repro-
> duced was made for a purpose other than that for which the
> performer's authorization was given, and
>
> (iii) where a fixation was permitted..., to reproduce any reproduc-
> tion of that fixation, if the reproduction being reproduced
> was made for a purpose other than one permitted..., and
>
> (c) to rent out a sound recording of it,
>
> and to authorize any such acts.

example), additional rights may be held by whoever organized that under-
taking.

The possibility exists that some of these neighbouring rights holders
may be dead. Sound recordings and performers' performances have a
copyright term of fifty years measured from the time of the performance
or fixation (not the death of the maker or performer).[4] In contrast, musical
works have the usual term of life of the author plus fifty years. If a composer/
lyricist/arranger of the work died less than fifty years ago, an estate will
represent that person and reap whatever revenues come along. But if the
composer/lyricist/arranger died more than fifty years ago, the work will
have become part of the public domain, where the ownership rights are
extinguished.

As complex as this may sound, it is the simplified version. Fortunately,
however, most of us don't need to know all the details, largely because
copyright collectives mediate most use and reuse of music. For example, the
Society of Composers, Authors and Music Publishers of Canada (SOCAN)
acts as an intermediary between its members and radio stations and other
commercial venues that pay it for the right to broadcast, or "perform,"

a repertoire of recordings in public. Radio stations also have licences with Re:Sound (formerly NRCC, the Neighbouring Rights Copyright Collective), which represents performers and sound recording makers. Music in coffee shops, skating rinks, airplanes, dance clubs, and answering services is similarly licensed.[5] While some areas of confusion might exist for business owners—for example, if they play a radio in their store out of one set of speakers they don't need a licence, but more elaborate setups will more likely attract SOCAN's attention—the average consumer doesn't usually have to worry, other than to ask whether the licence fees are appropriate and getting to the right people.

But let's look at things from the perspective of the composer. Let's say you have written a song. Apart from various users' rights exceptions, you hold the sole right to reproduction, performance, and broadcast of any substantial part of that song—and you hold the moral rights as well. You might seek a publisher's help in publicizing the song and negotiating terms with performers or labels; in this case, the publisher collects payments from collectives and passes on royalties to you. Through SOCAN, you can be paid when people perform your song or play a recording of it in public, according to a formula based on broadcast surveys. If your music is more likely to be played in a concert hall than on the radio, your rights will be the basis for your negotiation for performance or commission fees. If you want to encourage non-commercial use, you could arrange for a Creative Commons licence specifying the terms on which you want that to happen (for more on Creative Commons, see chapter 17).

If you are a performer, things are a little different. You have the right to fix your performance—which is to say, if someone tapes your live show and puts it on the Internet without your permission, that's infringement (though it could also quite well be welcome advertising, showing once again the awkward fit of law and reality). Once you have given permission for your performance to be fixed, you have no further say in its redeployment, but if you join Re:Sound you will be paid royalties when your recording is played in public.[6] Since 2012, performers also have moral rights just like composers and other authors—that is, they are entitled to attribution, and their work cannot be changed to the damage of their "honour or reputation."[7] The copyright term in performances—fifty years from their fixation—is shorter than the term for authored works, where the fifty-year period begins only after the author's death.

If you perform material whose rights are owned by others, those rights have to be cleared. In most performance situations, rights clearance is arranged by the promoter or the venue, not by the musicians. For example, a cover band in a bar works under that bar's performance rights licence with SOCAN, which represents the songwriter or the publisher. If you want to record a cover version of a work, you have to clear the "mechanical rights," which is the industry's term for reproduction rights. This is done on a case-by-case basis. It's not expensive: if the song is in the repertoire of CMRRA (Canadian Musical Reproduction Rights Agency Ltd.) or SODRAC (Society for Reproduction Rights of Authors, Composers and Publishers in Canada), and is less than five minutes long, the current rate is under $150 for

Section 2 of the Copyright Act defines a "sound recording" as "a recording, fixed in any material form, consisting of sounds, whether or not of a performance of a work, but excludes any soundtrack of a cinematographic work where it accompanies the cinematographic work."

Section 18(1) goes on to set out the exclusive rights of the owner of the sound recording:

the maker of a sound recording has a copyright in the sound recording, consisting of the sole right to do the following in relation to the sound recording or any substantial part thereof:
   (a) to publish it for the first time,
   (b) to reproduce it in any material form, and
   (c) to rent it out,
and to authorize any such acts.

0

one thousand CDs. If you can't find your composer through CMRRA or SODRAC, you can make arrangements independently. If the composer has been dead more than fifty years, you owe nothing for reproducing the work, as it is in the public domain.

Now let's say you want to sample a song from a sound recording—you want a second or two or three out of it, or you are laying some tracks on top of parts of it. If you know about the copyright substantiality requirement (chapter 3), you might think you don't need clearance. After all, if the source song is four minutes long, and you use three seconds, you've used 1.25 per cent of the song, which doesn't seem substantial. But the substantiality test isn't just about quantity. If you've taken the signature riff of that song, your use may be considered substantial. More importantly—because often samplers transform sounds well beyond recognizability—the industry practice that evolved in the wake of the 1991 Gilbert O'Sullivan/Biz Markie case in the United States (*Grand Upright v. Warner*) and that has been confirmed in *Bridgeport Music v. Dimension Films* (2005) is that rights have to be cleared for any amount or type of use.[8] The logic in these controversial U.S. cases has been that if it's worth sampling, it must be worth paying for. There is no case law on point in Canada, but we see no reason why substantiality and fair dealing should not apply to sound recordings as well as works. Canadian courts might decide in favour of samplers. Furthermore, if the use is not commercial, it may be protected by the new exception for non-commercial user-generated content in section 29.21. However, if you expect to distribute your recording in the United States, unlicensed sampling can be legally risky, and care should be taken. Sampling and DJ work are well-established art forms not well served by copyright law.

The Music Copyright Infringement Resource website (http://mcir.usc.edu) is a fabulous and even entertaining collection of recordings, cases, and commentary on U.S. music copyright.

Is downloading songs legal in Canada?

Section 80(1) of the Copyright Act indicates that the downloading (or other copying) of a song for a person's private use does not constitute infringement—if certain specific conditions are met.

To take advantage of this special exception, the copying has to be made to "an audio recording medium," and it has to be for the purpose of the "private use" of the person who actually makes the copy. That is, you cannot use this exception to make a copy for someone else. The exception is also not available if the copy is going to be sold, rented out, distributed, or communicated to the public by telecommunication or performance. Any distribution will take the copying out of the exception, even if it is not done for the purposes of trade or sales.

The question that remains somewhat unclear in Canadian law is whether putting a file in a shared directory ("uploading") constitutes authorization to reproduce it—if so, that act would be infringement on a right reserved to the owner of the copyright in the file. In the Federal Court case *BMG Canada v. Doe*, Justice von Finckenstein suggested that merely making a file available on a shared server did not infringe the owner's reproduction, distribution, or authorization rights. But this statement was not central to the case, which otherwise decided that record labels could not compel Internet service providers to identify subscribers suspected of copyright infringement. The Federal Court of Appeal affirmed this ruling, but said that the lower court should not have commented on the copyright issue.

Twenty-First-Century Music Rights

The structure of rights management in the Canadian music industry gets even more complicated when we take the Internet and digital music into consideration. The Internet has allowed more people access to more recorded music than ever before. The record store at your local mall, if there still is one, might carry Ke$ha and K'naan, but you can get access to their songs for free online, and venture further into formerly inaccessible field recordings, basement sessions, bootlegs, and remixes from artists famous and obscure, alive and long gone. Finding and sharing music online can lead to a sense of euphoria—with even more fun to be had sampling and mashing up the music.

These types of uses do not go unnoticed by the consumer electronics industry, which often bases its advertising campaigns on them—following Apple's 2001 campaign exhorting people to "Rip. Mix. Burn." Indeed, whole genres of creative activities are opening up because of the possibilities unleashed with new digital technologies. It is fair to say that the old distinctions between creators, producers, and consumers are imploding. Reproduction, recording, editing, and distribution have become cheap and easy enough that the mediating roles of record labels seem less necessary than they once were.

This is good news for many music lovers and some music creators, but it's bad news for the established music industry, or at least for the bigger labels that want to cling to the old ways of doing business. That, of course,

In the old days, when your album stiffed it was returned by retailers and cut out of the catalog, it no longer existed. Whereas today your work lives online forever, ready to catch fire if it's good and you can continue to draw fans to it. This is an opportunity. As is the streaming compensation model. You can continue to be compensated for decades if people care and listen. Listening is the key today. Hell, everybody knows "Gangnam Style"... It's the most popular YouTube video of all time, topping 840 million plays. It didn't enter public consciousness via radio, radio was last. And sales are puny compared to streams. This is the new world.

—Bob Lefsetz, "Fascinating Statistics," *The Lefsetz Letter*, www.lefsetz.com, 30 November 2012.

is where copyright comes into the picture. By increasing the sophistication of copying controls, lobbying for laws to layer on top of them, taking people to court, and otherwise intimidating consumers as best it can, the music industry has been trying to shut down or claim for itself the new capabilities of digital technologies. In Canada the music business has been lobbying heavily to stop downloading, which is ironic considering that the industry provided the impetus for the mechanism that makes downloading legal: the private copying levy by which Canadians pay a premium on blank media on the presumption that they will use them to copy music.

While the recording industry purports to speak on their behalf, most musicians have little time or money for lobbying—nor are they eager to bite the hand that sometimes feeds them. Many artists are resentful about downloading, but many also see the Internet as a way of asserting both professional and political independence from the large trade groups. In the decade of copyright ferment that ultimately resulted in Bill C-11, Steven Page of the Barenaked Ladies, along with other prominent Canadian musicians, established the Canadian Music Creators Coalition, which spoke out against suing consumers and locking up digital content and called instead for more direct funding of artists. "Record companies and music publishers are not our enemies," the CMCC said on its website. "But let's be clear: lobbyists for major labels are looking out for their shareholders, and seldom speak for Canadian artists. Legislative proposals that would facilitate lawsuits against our fans or increase the labels' control over the enjoyment of music are made not in our names, but on behalf of the labels' foreign parent companies."

Practical alternatives to the big label model are clearly needed. Musicians may want to be independent from the big labels, but the everyone-for-themselves model of selling CDs through personal websites is ineffective and inefficient for most. When Charlie Angus of the NDP proposed what was promptly termed an "iPod tax" in 2010, he was laughed out of the House.[9] We're not so sure it was a crazy idea. According to a 2011 report by Columbia University's American Assembly, 46 per cent of U.S. adults have downloaded unauthorized music, TV shows, or movies, and the figure is 70 per cent for those under thirty. The report also says that those who do unauthorized downloading also do 30 per cent more authorized downloading, but that conclusion has been challenged by other studies.[10] Another U.S. study from 2011 found that Napster and peer-to-peer sharing (P2P) have not resulted in a reduction in new recorded music.[11]

Some consumer rights advocates would say that, in that case, there is no problem. But the study didn't prove that musicians are making a decent wage. Blog comments like "Recordings should be promotional & the live show is where they should be making thier [sic] money. Working for it"[12] come from people who would probably not take kindly to others telling them that the products or spreadsheets they generate in their jobs should just be promotional. Such people may also have no idea of how little most club and tour gigs pay: many talented and committed musicians are lucky to get gas covered and in fact depend on CD sales (not to mention other employment) to bring in enough to survive.

It still seems to us, therefore, that a levy on Internet service provider fees could be appropriate, even if after *SOCAN v. CAIP* ISPs do not have to worry about legal liability for downloading on their servers. Collective licensing of P2P could clarify a practice that is already prominent, and presumably avoid the digital locks that record companies might otherwise be inclined to impose on purchased materials. It might also avoid, we would suggest, some of the consumer ire and cross-border shopping that a levy on digital recording devices could lead to.

That said, past experience might lead many Canadians to doubt the appropriateness of a collective licensing solution to Internet music issues. On the creator side, a levy system could present as many problems of fairness or accuracy when it comes to survey techniques or allotment mechanisms as there are today with the big label system. For consumers, the pitfalls are multiple (see the discussion of the levy on blank media in chapter 6). On the decentralized Internet, even more than on previously licensed media, we have the added problem that a large amount of the music circulating is either already in the public domain, orphaned (we can't name or locate the rights holders), never intended to be commercial, or already paid for

The Supreme Court on Music, 2012

Four of the five copyright cases from the Supreme Court in 2012 dealt with some aspect of music:

1) In the related cases of *ESA v. SOCAN* and *Rogers v. SOCAN*, the court distinguished between downloading and streaming of files containing musical works: "A download is the transmission over the Internet of a file containing data, such as a sound recording of a musical work, that gives the user a permanent copy of the file to keep as his or her own. A limited download allows the copy to be used as long as the user's subscription is paid up. A stream is a transmission of data that allows the user to listen to or view the content transmitted at the time of the transmission, resulting only in a temporary copy of the file on the user's hard drive" (*Rogers*, para. 1). When a user *downloads* a file, it constitutes a reproduction. But when a file is *streamed*, the court said, it does not constitute a reproduction, but rather a communication to the public by telecommunication, which is historically related to the owner's exclusive right to perform the work in public. So these rulings distinguish rights implicated by each action.

2) In *Rogers v. SOCAN*, the court also clarified that the streaming of files that are available to users through online music services constitutes a communication by telecommunication of that work *to the public*. Rogers had unsuccessfully argued that these communications were not to the public because they were requested by different users at different times. The court disagreed, giving "to the public" a broader interpretation in the case of Internet transmissions.

3) In *SOCAN v. Bell*, the court held that previews of songs provided by online music service provides constitute "research" for purposes of fair dealing and thus are not eligible for a tariff.

4) In *Re:Sound v. Motion Picture Theatre Associations of Canada*, the court clarified the definition of "sound recording" from section 2 of the Act. The definition excludes soundtracks that accompany a film. When a soundtrack "accompanies" the film (as in the case of a broadcast or in a theatre), it does not technically constitute a "sound recording" (and does not require separate compensation owed to owners of sound recordings). But if the soundtrack is extracted from the film, presented separately (as in the case of a CD or recording of just the music) then it is within the definition of a sound recording.

by individual contract (iTunes, for example). Levies would probably be inflated without due recognition of these conditions. Finally, there is something inherently distasteful to many in the idea of switching the default conception of the Internet from an open-ended information commons to an enclosed and metered market.

Canada's cautious approach in responding to the so-called music piracy crisis has been productive, and we expect new proposals to emerge in the future. We can agree that the current system doesn't work, but we need to take our time to see what its replacement might be.

10. DIGITAL MEDIA

I n the era of personal websites, blogs, cloud computing, user-generated content, and other forms of Internet communication, the old distinctions between publisher and consumer, or content provider and user, have broken down. The author of material on the Internet may also be its publisher, and its reader may soon become a co-author because of some interactive feature that encourages instant transformation. One effect of this situation is that people who know nothing about copyright law, and have never thought of themselves as owners of copyright works, are now routinely able to produce, reproduce, transform, and distribute works. These works will then be used by others who will repeat this process of reuse and transformation.

Consent and Terms of Use

Copyright law is supposed to be technologically neutral: it should treat material on the Internet the same way as it treats any other material. If the

material meets the requirements for copyright to subsist (s
is automatically copyrighted. By making materials available
owners are not foregoing their copyright interests, nor are t
anyone to go out and reprint and distribute the work on a c⌐⌐⌐⌐⌐⌐ ⌐⌐⌐ ⌐⌐⌐

If you are creating original content that you upload to a website, you can clarify what use of the material you consider appropriate. You can state terms of use in your own words—by entering a line such as "Non-commercial use of material on this site freely licensed," or "All rights reserved. Copyright 2013 Joe Dragoneater." These are not binding contracts, but they do advise users of your preferences.[1] For many purposes, the Creative Commons licensing system is useful (see box on page 139 and chapter 17): it offers an array of choices in licensing terms, such as permission for non-commercial use only, or permission for reproduction as long as no changes are made to the original.[2] If users want to make a non-permitted use, they can contact you and negotiate, which means that Creative Commons licences can be suitable even for people who seek commercial outlets for their work. Visual artists and photographers can also turn to technological protections: they can put work online in low resolution or layered with watermarks—suitable for browsing and recognition, but not for quality reproduction. Musicians can enable streaming without enabling the ability to download.

But users' rights also obtain on the Internet: the Act's exceptions—such as fair dealing and the new exception for user-generated content, discussed below in this chapter—can justify various uses and reuses of the work. Beyond that, there is implied consent. If an owner of copyright material uploads works to a server enabling public access without stated limitation, that uploading is a form of consent for a reasonable range of uses. Anyone who is able to upload content to a website also understands that computer users have access to an interface that includes capabilities like save, copy, paste, print, export, and select all. The vitality of the Internet depends on general acceptance of diverse uses and reuses. Emerging practices suggest that Internet users can expect refusal of consent for such uses to be explicitly stated or, more likely, demonstrated through encryption, password protection, or the application of some other technological measure to restrict what they can

The Technological Neutrality Principle

In *ESA v. SOCAN* (2012), Justices Abella and Moldaver writing for the majority asserted the importance of "technological neutrality" in interpreting the Copyright Act. The Copyright Board had granted the Society of Composers, Authors and Music Publishers of Canada (SOCAN) a tariff based on "communication to the public by telecommunication" where video games containing musical works were distributed over the Internet. The Entertainment Software Association (ESA) objected to the additional cost, which would apply to copies sold by download but not to those purchased in material form. The court agreed with ESA that the Board's ruling violated the principle of technological neutrality:

> In our view, the Board's conclusion that a separate, "communication" tariff applied to downloads of musical works violates the principle of technological neutrality, which requires that the *Copyright Act* apply equally between traditional and more technologically advanced forms of the same media. . . . In our view, there is no practical difference between buying a durable copy of the work in a store, receiving a copy in the mail, or downloading an identical copy using the Internet. The Internet is simply a technological taxi that delivers a durable copy of the same work to the end user. (para. 5)

While the case arose in the context of online video games containing musical works, the ruling is stated in broad terms and may be viewed as a guideline for many other fact situations:

> The principle of technological neutrality requires that, absent evidence of Parliamentary intent to the contrary, we interpret the *Copyright Act* in a way that avoids imposing an additional layer of protections and fees based solely on the *method of delivery* of the work to the end user. To do otherwise would effectively impose a gratuitous cost for the use of more efficient, Internet-based technologies. (para. 9)

do with the file. Only a few years ago it would have been difficult for the non-technically minded to apply such protections; today it is simple and routine. For example, when you save a PDF in Adobe Acrobat, you can easily disable users from applying the copy function. Thus, even if the web page in question does not say the magic words "The owner of the copyright in this work hereby gives you consent to download, print out, distribute, and use the copy and paste function to extract portions of this material for reuse," that sentiment is implied if it is not prevented and no restriction is stated. After all, that is how people use the Internet.

When it comes to reusing works owned by others directly on your own website as a substitute for work you would otherwise have had to do yourself or pay for, the implied consent theory described above starts to lose its strength. Reproducing a photograph, drawing, or piece of writing you found online (or a substantial portion of the work in question) for your newsletter or zine without permission could be infringement. The usual rules about what constitutes infringement apply on the Internet as everywhere else (see chapters 3 and 4): you have to review, for example, whether the reproduction is substantial and whether an exclusive owner's right is triggered. If the answer to these questions is yes, consider contacting the person who took the photo, drew the cartoon, or wrote the poem to see how much they would charge to license the work for your site. There's sometimes a tendency to think that rights holders will be rapacious, but the opposite is usually true. Consider the millions of photos on Flickr: many of those photographers would be glad to have their work used for a modest fee or even just for a polite acknowledgement.

However, we reiterate that fair dealing is as available online as owners' rights are—even if you would not know that from various copyright notices on commercial sites. The *Globe and Mail* certainly tries to obliterate it:

> The Globe and/or its licensors grants you a limited non-exclusive, non-transferable license to use and display on your computer or other electronic access device, the Content and Services for your own personal, private and non-commercial use only, provided that

you do not modify the Content and that you maintain all copyright and other proprietary notices. Except as provided herein, you agree not to reproduce, make derivative works of, retransmit, distribute, sell, publish, communicate, broadcast or otherwise make available any of the Content obtained through a Globe Property or any of the Services, including without limitation, by caching, framing or similar means, without the prior written consent of the respective copyright owner of such Content.[3]

This statement renders itself rather absurd, given the routine caching functions of any personal computer. But the big vacuum here is fair dealing: if a website's use policy tells you that you can't quote from it, or print it out for a scrapbook or research file, remember your users' rights.

Another example: CBC.ca's permissions FAQs answer the question "Do I need permission to use CBC.ca content?" with the blunt statement "Yes. Any content . . . found on CBC.ca can only be reused elsewhere with the permission of CBC."[4] But the CBC cannot simply abolish a large portion of the Copyright Act by fiat. It is particularly ironic for large news organizations to try to impose such onerous terms and conditions on their readers, since the press is itself a major beneficiary and claimant of fair dealing for purposes including news reporting.

Digital Rights Management

The limitations imposed by copyright owners on what you can do with digital material are not always readily apparent. Often, and sometimes even without your knowing it, uses are controlled or prevented by software. You try to copy a scene from a DVD into your movie editing software for a class presentation and find that you can't. You switch to Linux and lose access to your iTunes library. You see a photograph in a proprietary database, but you can't copy it—even though the copyright has expired and the work is in the public domain. You discover that an electronic document that arrived two weeks ago has disappeared without a trace from your hard drive. You

Creative Commons is easy to use for both makers and users. *From "Radical Extremism" to "Balanced Copyright": Canadian Copyright and the Digital Agenda*, a collection of essays edited by Michael Geist—while it was published by a commercial publisher and is available for sale in the regular book market—is also available online under the terms of a Creative Commons licence. To obtain it online, you take the following steps.

- Go to the publishers' web page for the book at www.irwinlaw.com/store/product/666/from--radical-extremism--to--balanced-copyright
- Select any one of the essays from the table of contents.
- Scroll down until you see a box labelled "Creative Commons Legal Code Attribution-NonCommercial-NoDerivs 2.0 Canada." While the text seems to be full of legal jargon, if you read it section by section it is understandable. (Also, you can go to http://creativecommons.org/licenses/by-nc-nd/2.0/ca for a more streamlined version.)
- At the end of this text, there is another small box labelled "I agree to the terms and conditions listed above." If you indicate your agreement to the terms, you are granted "a worldwide, royalty-free, non-exclusive, perpetual (for the duration of the applicable copyright) licence" to exercise certain enumerated rights.
- If you check that box, you can then download the PDF of the article. Notice that the PDF file is an exact reproduction of the page in the book as published, with the page numbering intact. You are also able to use features such as save, copy and paste, print, and forward with respect to the article.

Clearly, the editor and the authors of articles in this collection want you to use their work. They want you to read it, they want you to share it with others, they want you to cite the articles in your own transformative work, and they probably want you to ask your librarian to purchase the hard copy of the book. Creative Commons can facilitate all of these actions.

Can I use cartoons from the Internet in my PowerPoint presentations?

It would depend on the facts and context. Such a use might be fair deal-ing, but probably only if it genuinely engages in criticism or review of the given cartoons, and if you can justify that you had to use those particular cartoons. If you are making your presentation in an educational institution, section 29.4(1) of the Copyright Act allows projection of copyright materi-als, or it could be fair dealing. But "educational institution" is defined in the Act as a non-profit government-recognized institution, which does not cover sales conferences, some private schools, or public lectures.

You might decide to go ahead based on the rationale that it is practically impossible to determine the author or copyright owner of some much-circulated but unlabelled images on the Internet—and it is hard to imagine that the Copyright Board would want to bother with licensing one-time use on behalf of such unlocatable owners. You might suppose that the potential for liability is negligible (probably true if your presentation is ephemeral), or that the creators wouldn't mind in any event (may or may not be true). You might argue that it's an established practice in your industry or sector—meaning that the practice is generally accepted in like situations—in which case you are getting close to an implied consent argument.

discover that a digital resource bought by your library a few years ago can no longer be accessed.

Welcome to the world of digital rights management, known, not always fondly, as DRM. You often hear the acronyms TPMs (technological protection measures) and RMI (rights management information) as well; these are subsets of the broader practice of DRM.

DRM is a mechanism by which owners or vendors of digital intellectual goods can control access to and use of the materials they make public or sell to consumers. As we saw above, it can help creators or businesses publicize their work online without giving it away. It can also be used less benevolently to assert claims or regulate practices beyond the bounds of copyright.

An economics of information approach illuminates the effects of DRM. Intellectual or information goods, as we saw in chapter 1, are public goods in the sense that they have no built-in exclusion mechanisms and they are non-rival in their consumption; these characteristics are particularly evident now that they can be embodied digitally. Legal and technological exclusion mechanisms make public goods into private goods.

In recent years, exclusion mechanisms have made many applications of digital technologies commercially and practically useful. Software encryption can ensure the security of your bank account and the privacy of your email. It allows online businesses to set passwords for subscription access, and it can guarantee the authenticity of electronically transmitted materials. These are useful tools and we all rely upon them. But exclusion mechanisms can also be used as ways of fencing in or enclosing information spaces that in their analog form were open for common use. For example, a commercial enterprise might digitize old public domain books—and then make them available by subscription only.[5]

The book, in fact, can help us understand the effects of DRM. It used to be that copying a book required longhand transcription or retypesetting. Then we moved into the eras of photography, photocopying, and scanning. The copying of paper originals has become easier, but it is still expensive and time-consuming, and there is always a noticeable loss of quality. These constraints limit copyright infringement, which may make some rights holders look wistfully at the old technology of the book.

But we might also consider some of the qualities of books that are enabling for users—or, as we used to call them, readers. You can read books from beginning to end, in bits and pieces, or even backwards. You can read a book in secret or in public, in silence or out loud. You can quote from it. You can place bookmarks in a book, write in the margins, or even tear out the pages and tape them to the wall. You can lend a book to a friend, and libraries—established by the mid-nineteenth century throughout North America—were designed to exploit this very capacity. You can also keep a book for a very long time. As long as you know the alphabet and the language and have taken care of the book, you can read it centuries after it was written. Finally, if you don't want a book any more, you can give it away or sell it.[6]

This quick summary reveals that we can do a number of things with purchased books that we may not be able to do with purchased digital files. There is nothing inherently limiting about digital technologies; in fact, quite the contrary is true. But while digital media have the potential for expandability, unlimited usability, portability, sharing, and preservation, the prevalence of new forms of DRM brings new types of limitations.

For instance, if an e-book didn't have DRM imposed on it, you would be able to copy it, and your copy would be cheaper and better than most copies of paper books. But it does have DRM, and even if you know how to work around that limitation and make a copy, another level of DRM might prevent you from doing anything with that copy. It doesn't matter what your goal is: your book might be an old work that is now in the public domain, or you could be practising fair dealing, but the DRM would not discriminate.

In any case, quite likely you never did hold all of the usual legal rights associated with ownership when you acquired the e-book. A careful look at the terms of use shows that you really only purchased a limited licence to do certain things with it. Under typical contracts for electronic media, you give up your rights to sell the material, or even lend it, and you may give up the ability to change it into a different format or play it on a device other than the one preferred by the vendor. For the vendor, contractual provisions often reserve rights that the Copyright Act does not—rights that historically belonged to purchasers of copyright materials. These reserved rights, sometimes known as "paracopyright," are found in licences governing the use of computer programs, video games, and iTunes songs, for instance.[7]

While vendors justify such restrictive practices by the need to prevent copyright infringement, they are actually unilaterally changing the nature of the market by limiting purchasers' uses of the goods and thus, presumably, increasing sales. Vendors are developing call-home mechanisms to track the amount and nature of use, and will increasingly base pricing or marketing strategies on the information gleaned in this way. Economists refer to this practice as "price discrimination": the idea is that vendor and consumer both benefit from a price based on the amount of use. But to enforce usage restrictions, along with other provisions in the contract, call-home mechanisms may be invading users' privacy.

Ironically, DRM is not proving effective against industrial-scale piracy. The biggest leak in the exclusion mechanism for movies, for example, appears to be people with video cameras in cinemas, not people with advanced de-encryption skills. What DRM is doing is frustrating law-abiding consumers' use of material that they have legally acquired. This makes it all the more ironic that in 2012, Canada finally capitulated to persistent U.S. pressure and added new provisions to the Copyright Act that prevent circumvention of DRM subject to some very limited exceptions. (For more information, see chapter 8.)

Internet Linking

The Internet offers one alternative to reproduction not available so easily in other media: linking. Linking to someone else's website is a perfectly legitimate practice. Using hypertext links is the essence of the World Wide Web, and it seems strange that anyone would object to having external links coming to their site. Yet owners of websites do raise such objections, and as a result linking practices can raise certain other legal problems. Linking practices range through three different types: direct linking, deep linking, and framing.

Direct linking is the simplest form of using a hypertext link. It's hard to imagine anyone complaining about getting an external link sent over from another site. After all, as they say, the Internet is all about eyeballs. But some websites do not want the wrong kind of eyeballs, and they try to discourage external links that are created without their permission. For example, the Terms and Conditions page on the Exxon/Mobil website advises, "You may not link to this site without prior written permission from ExxonMobil."[8] The web page for the New York Stock Exchange says, "NYSE Euronext prohibits caching, unauthorized hypertext links by others to the NYSE Euronext Website and the framing of any Content available on its NYSE

Euronext Website."⁹ These kinds of restrictions are blatantly unreasonable, and if you come across such statements you should not give them much concern, at least not from a copyright liability point of view.

The practice of *deep linking* presents slightly different issues. According to Wikipedia, "deep linking is making a hyperlink that points to a specific page or image on a website, instead of that website's main or home page. Court cases against deep linking have gone both ways in various countries."¹⁰ Linking to an internal page in a website is much like going to a card catalogue at a library, finding the book on the shelf, and looking through its index for the material you are seeking: you don't have to start reading at page 1 to find what you want. Deep linking has been the subject of controversy because some website owners want visitors to enter through the front door and only through the front door. They do not want other sites to send visitors into the interior of their site without passing through the portal of the home page. This may be because there is advertising on the home page, because they want readers to follow a certain order of presentation, or because they just want to maintain more control over the situation. As a copyright issue, demands to refrain from deep linking should not carry much weight. In any event, a web designer should be able to redirect users back to the appropriate web page if someone tries to enter through a restricted page.

Framing is a very different practice than linking or deep linking because you are actually bringing the external web page into a portion of your presentation, as a framed screen. Depending on how you set the frames up, the boundary between the external materials and your own may or may not be entirely clear. From a copyright point of view, it is hard to argue that there is any direct infringement because you are not really reproducing the external site on your own page. But it could appear that way to the end-user. Be careful that you are not setting your frames up in a way that misrepresents exactly which content is yours. While copyright issues seem remote in this regard, other legal theories such as misrepresentation or passing off could be relevant. These restrictions on different types of linking have generated litigation in other countries, and there have been recent cases in Canada treating the legal implications of linking.[11]

User-Generated Content

We end this chapter with a discussion of user-generated content, a growing and important new practice in the area of digital media that warrants some special attention. "User-generated content (UGC)" is coming into use both as a catch-all term for a range of commonly understood creative practices, and as a legal term of art in copyright.[12]

In 2012, Bill C-11 added section 29.21 to the Copyright Act under the heading "Non-commercial User-generated Content":

(1) It is not an infringement of copyright for an individual to use an existing work or other subject-matter or copy of one, which has been published or otherwise made available to the public, in the creation of a new work or other subject-matter in which copyright subsists and for the individual—or, with the individual's authorization, a member of their household—to use the new work or other subject-matter or to authorize an intermediary to disseminate it, if

   (a) the use of, or the authorization to disseminate, the new work or other subject-matter is done solely for non-commercial purposes;

   (b) the source—and, if given in the source, the name of the author, performer, maker or broadcaster—of the existing work or other subject-matter or copy of it are mentioned, if it is reasonable in the circumstances to do so;

   (c) the individual had reasonable grounds to believe that the existing work or other subject-matter or copy of it, as the case may be, was not infringing copyright; and

   (d) the use of, or the authorization to disseminate, the new work or other subject-matter does not have a substantial adverse effect,

financial or otherwise, on the exploitation or potential exploitation of the existing work or other subject-matter—or copy of
it—or on an existing or potential market for it, including that
the new work or other subject-matter is not a substitute for the
existing one.

(2) The following definitions apply in subsection (1).

"intermediary" means a person or entity who regularly provides
space or means for works or other subject-matter to be enjoyed by
the public.

"use" means to do anything that by this Act the owner of the copyright has the sole right to do, other than the right to authorize anything.

The exception in section 29.21 is very broad: it applies not just to works
but to all copyright subject matter including sound recordings. And it is
not limited to making copies, as the term "use" applies to all of the owner's
exclusive rights other than the authorization rights. So the new exception
applies to public performances, translations, adaptations, and communications to the public of works and sound recordings. It could apply to fan vids
and fan fiction, for example, but also to many sorts of collage or remix or
layering of materials, use of data sets, and programming efforts (see Table
11). Also, there is no limitation to online uses in the text of the exception, so
it could be applicable to audio and print reproductions.

But there are also substantial limitations in paragraph (a) through (d).
First, the content being used must have been "published or otherwise made
available to the public." Second, the use must be "solely for non-commercial purposes." This is a bright-line rule, as lawyers say: there is no fudging
it. If there are commercial purposes, the exception will not be available.
As a practical matter this may be a difficult distinction, as the commercial/
non-commercial nature of use might shift over time. What happens if the
UGC begins as a wholly non-commercial project, such as a school project or a hobby activity, and it subsequently enjoys a measure of success?
Would the previously attached UGC exception remain intact, or would it

Table 11. Domains of User-Generated Content

A. Creative Content	B. Small-Scale Tools	C. Collaborations
• Content created, developed, captured and put on display by an individual on an online platform • Content generated by individuals or small groups (not within virtual worlds or gaming platforms) • More specifically, platforms such as YouTube, Flickr, Twitter, and Facebook • UGC where an individual (or small non-regulated group) is in control of creation of content and uploading it for delivery on a platform	• Tools, modifications, and applications created by a user or group of users • Game modifications (mods), or add-ons • Mods, objects, or tools created for virtual worlds such as Second Life • User-developed applications and tools for mobile devices, such as iPhone or Android systems	• Collectively authored materials shared by a self-regulating group of contributors • Includes open source and free/libre software • Wikis, such as Wikipedia • Government data sets made freely available

Source: Trosow et al, "Mobilizing User-Generated Content for Canada's Digital Advantage."

then be nullified? We will have to watch how the practice develops in this area.

The final condition, that the use cannot have "a substantial adverse effect," may also be problematic. The effect does not have to be financial, the exploitation can be actual or potential, and the market can be existing or potential. This could in practice be a major constraint on the exception. Of course, as with all particular exceptions in the Act, we note that this one, too, is additive to fair dealing as a safe harbour: if a failure to meet one of the conditions disqualifies the UGC from the exception, fair dealing might still be available.[13]

There is no question that the phenomenon of UGC has had a somewhat destabilizing effect on existing business models. As Debora Halbert has written:

> Where once there existed the relatively stable world of the culture industry in which concentrated control over film, music, literature, and art was easy, the technology of modernity has shifted control into the hands of consumers of culture. Stable control over the culture industry was possible because commodity culture de-skills people as creators, in the same way that industrialization de-skilled the artisan and crafts-person while turning them into fodder for the industrial machine.[14]

The relationship between UGC and copyright in Canada presents a paradox. On the one hand, creative practices have informally evolved in spite of copyright restrictions, and as these practices have become widespread and accepted, they have become an impetus for reform in a user-oriented direction. One the other hand, the new protections for DRM will limit and impede the ability of creators to actively engage with copyright materials. We think that Canada has the potential to become a leader in UGC, and that the exception in section 29.21 will have positive economic, social, and cultural effects. However, those effects will to some extent be dependent on how the DRM rules become enforced and regulated.

11. FILM, VIDEO, AND PHOTOGRAPHY

The world of photographic images, both still and moving, has much changed in recent years. The proliferation of cheap new camera technology, easy-to-use editing software, and free distribution mechanisms (YouTube, Instagram, and Flickr being the most famous at the moment) has increased the number of amateur photographers and filmmakers exponentially. The Internet and demand for material for mobile digital devices and video games have enlarged the market for video and photos, so there is a place for more and more small entrepreneurs. And the Internet has opened the door to new business models that may not entirely depend on "all rights reserved": some video makers might make work freely available online as a resumé, giving it away in order to attract more jobs. Like music, it's a bit of a confusing environment, in which filmmaking is less and less synonymous with the film industry as we used to know it, and in which the line between amateur and professional photographers has become blurred.

In terms of the law, one thing has been clarified since the first edition of this book in 2007: whereas photography used to have special rules about ownership, it is now treated like all other works. In this chapter, we will treat it alongside film.

Although owners' rights are of course a concern for professional image-makers (chapters 4 and 10 will be of particular interest in this regard), we focus in this chapter on users' rights, because they seem to present especially persistent challenges.

Users' Rights

One pressing intellectual property problem for filmmakers and photographers is that urban spaces are overflowing with trademark or copyright material—any shot is likely to contain a corporate logo or a fragment of recorded music. Even inside shoots are risky: clothing, background posters, and TV constantly present proprietary images to the camera. If you are working on a large film project, the prospect for product placement fees can work to your advantage, even though you might be expected to give up some artistic control. But if you are working with a small audience in mind, or trying to maintain artistic independence, companies might try to collect money from you for these glimpses of proprietary images.

Luckily, there are quite a number of defences to such claims. First, a lot of the concern about logos comes from trademark law, not copyright law. There is little basis in trademark law for demands from trademark owners for royalties on images of a logo. The courts have demonstrated repeatedly that unless there is a strong likelihood of consumer confusion, reproducing or mimicking a trademark does not constitute trademark infringement. In other words, a finding of trademark infringement requires that the mark be used in a related line of trade. In two Canadian Supreme Court cases, a fast-food restaurant chain, Barbie's, and a clothing store, Boutique Cliquot, were allowed to keep their names because they were not selling the same thing as the original trademark owner.[1] These decisions suggest that under trademark law, if you're not selling hamburgers, you can reproduce the "Big M"—and if you're not selling leatherwear or sweatshirts, you can reproduce the Roots logo. The owners of Budweiser might not be happy to see Denzel Washington guzzling their product at the wheel of flight 227 to Atlanta in the movie *Flight*, but they are unlikely to be able to block this use in court.[2]

Since commercial logos are not only trademarks—they are also artistic works in which copyright subsists—you can also look to the Copyright Act for some relief from the permissions burden. There are several reasons why reproducing copyright artwork without permission might not constitute infringement: substantiality, fair dealing, and specific copyright exceptions.

Regarding substantiality, see our discussion in chapter 4 about the threshold requirement. Many fleeting bits of music and other copyright material constitute non-substantial reproduction and thus might not even implicate section 3 rights.

Furthermore, with *Alberta v. Access Copyright* (2012) confirming *CCH v. Law Society of Upper Canada*'s assertion that the fair dealing categories ought to receive a "large and liberal interpretation," we can plausibly argue that a great number of photographic or documentary uses constitute criticism, news reporting, parody, or satire. If you think yours does, try out the six fairness tests from the *CCH* case (see chapter 5). Remember that the fact that your work may be commercial does not disqualify it from fair dealing. And the court even said in *CCH* that use of an entire image might be permissible fair dealing: this provision can be broader than you might think.

In addition to the substantiality test and fair dealing, three specific exceptions in the Act may shorten your list of rights to clear: the incidental inclusion, buildings and public art, and non-commercial user-generated content exceptions.

The incidental inclusion exception in section 30.7 states:

It is not an infringement of copyright to incidentally and not deliberately (*a*) include a work or other subject-matter in another work or other subject-matter; or

(*b*) do any act in relation to a work or other subject-matter that is incidentally and not deliberately included in another work or other subject-matter.

This exception has never been tested in court, and proving "incidental" inclusion in an edited film might be quite difficult. Still, it would seem to have been devised just for the serendipitous documentary situation in which the mental-hospital patient breaks out into song, or the firefighters are sitting around watching *The Sound of Music*.

The buildings and public art exception is stated in section 32.2(1):

It is not an infringement of copyright ...

(*b*) for any person to reproduce, in a painting, drawing, engraving, photograph or cinematographic work

(i) an architectural work, provided the copy is not in the nature of an architectural drawing or plan, or

(ii) a sculpture or work of artistic craftsmanship or a cast or model of a sculpture or work of artistic craftsmanship, that is permanently situated in a public place or building.

This provision should allow filmmakers and photographers to ignore permission demands from owners of buildings and public art—unless permission is needed to get access to private property.[3]

Finally, there is the new exception for non-commercial user-generated content enacted in 2012. Section 29.21 of the Copyright Act now provides that existing work may be incorporated into new works without permission or payment. This exception is not uniquely applicable to video, but it arose out of consumer outrage over certain notorious complaints lodged by corporate rights holders when people posted YouTube videos using music without clearance, so it is associated with burgeoning amateur video practice. The exception can only be invoked in limited circumstances: it must be done for non-commercial purposes; acknowledgment must be made of the source "if it is reasonable in the circumstances to do so"; and the exception only applies if there is no effect on the "existing or potential market" of the original (for more thorough analysis, see chapter 10). In other words, although this exception is much narrower than fair dealing, it may be useful not only for amateurs but also for student and emerging photographers and filmmakers, who can operate under its umbrella until such time as they seek to sell their work.

Reality Check

Users' rights as we have outlined them above ought to reduce permissions hassles and costs a great deal. However, for those seeking broadcast, distribution, or sometimes exhibition, insurance companies often demand permissions even when they may not seem necessary. Documentary filmmaker Kevin McMahon explains the experience of working with errors and omissions lawyers this way:

> Anything we sell to a broadcaster has to be insured. They demand E&O insurance to protect them in case they're sued. The way it works is that our lawyer will go through a production and they'll flag it: they'll say this, this, this, and this are a problem. Change 'em. If you decide to stand your ground, then the ultimate decision is made by a lawyer working for the insurance company. And so I might want something to be in, and even a commissioning editor at a broadcasting company might want something to be in, but both of us are overruled by, first, the corporate policy of the broadcaster that you must have this insurance, and finally by a lawyer in the employ of an insurance company.[4]

Countless filmmakers say the same thing: it is E&O insurers who make the law. In this environment, filmmakers have to design their films and schedules to minimize permissions (which may amount to a misrepresentation of the commercialized world), and they have to budget time and money for

One day in 2005 an employee of CKY, the Winnipeg CTV network affiliate, called the filmmakers of a collective called L'Atelier National du Manitoba and offered them three carloads of videotapes that the station was about to throw away. The product of this gold mine was a film about the demise of the Winnipeg Jets NHL franchise. *Death by Popcorn* attracted attention in Winnipeg hockey and art circles, but it didn't become nationally famous until CTV decided to go after its makers for copyright infringement. The *Globe and Mail*, CBC, and *Winnipeg Free Press* pounced on the story, and public outcry emerged in letters to the editor and blogs. L'Atelier member Matthew Rankin recollects:

> And then out of the blue we get an e-mail from the general manager of CKY, and he was like oh jeez, I didn't know that this was such a big deal, I think we can meet and talk about this. So we went and talked to him, and it was really bizarre because he was very friendly and he bought us lunch. And he apologized. He said I'm sorry that I sent the operations manager after you. This is my responsibility, we were just worried about the liability if the NHL came after us . . . Just disregard that letter, you don't have to sign or anything. You can show this movie at film festivals, that's no problem, because that's definitely not a commercial type vehicle, and I hope we can be friends. You should come and pitch us documentary ideas, and this kind of thing. It was the most amazing thing because literally the last time I was at CKY they really treated us like we were the scum of the earth, and now their general manager is begging for forgiveness. I really feel that it can be kind of an inspiring story to people who run into this kind of problem. I can't help but feel very cynical about it in a way, because if they didn't get a bunch of bad press then I would have had to sign something saying that I broke the law. And I could've been sued, and all of this. So a little bad press went a long way with them and they ended up dropping the thing.

Source: Matthew Rankin, Winnipeg, interview with Kirsty Robertson.

the permissions that do remain. Often, in the end, they can only afford permissions for limited territories or time periods, which dooms their work to obsolescence. In what the Documentary Organization of Canada (DOC) calls a growing "clearance culture," even student film festivals are now asking for copyright clearance at the time of initial application.[5]

The good news is that in both Canada and the United States, there seems to be increasing pushback by filmmakers and public interest advocates. An increasing range of films have been making use of fair use or fair dealing, often in combination with clearance for some material. In Canada, Brett Gaylor's *RiP! A Remix Manifesto* (2008) was spangled with uncleared clips and made rather a big deal of its likelihood of attracting lawsuits, but it never did. Other recent Canadian films that relied partly on fair dealing (*Reel Injun*, *Shameless: The ART of Disability*, *The Corporation*) have also remained unhassled.[6] Some of these films were insured by public broadcasters, which had a higher risk tolerance. The French film *Logorama* (2009) didn't clear rights either: it was made up entirely of 2,500 images of logos, and although it won the Prix Kodak at Cannes and an Academy Award, no one ever sued. Other more widely distributed films have not sought clearances and have not triggered lawsuits: *Super Size Me*, the documentary film about a man who lived entirely on McDonald's food for a month, is a prominent example. A film such as this is protected by the lack of money behind it, and the public outcry that would arise should McDonald's sue. The cost to the company's image would not be worth anything they might gain financially. Many more U.S. examples can be found on the Fair Use site of the Center for Social Media (www.centerforsocialmedia.org)—if lawyers suggest the U.S. environment is more dangerous, direct them to this site.

The *CCH* decision not only defines fair dealing broadly, but also offers recognition of the importance of norms of particular trades and industries. The fact that both Canadian and U.S. documentary film associations have developed statements of best practices for fair dealing or fair use is an important foundation for invocation of those rights, because it shows the existence of such norms among documentary filmmakers—not to mention that the documents provide useful guidance.[7] Photographers have not devised such a statement, but for art photography, at least, one could point to a substantial contemporary art practice that incorporates copyright or trademark material, going back to Andy Warhol and beyond (see chapter 12).

A final challenge for documentary filmmakers that must be acknowledged is the high cost of footage and photograph clearance. Private and

In 2000 I did a personal documentary called *Sea in the Blood*, and I wanted to use a line of text from Joni Mitchell, which goes "The wind is in from Africa/ last night, I couldn't sleep." There was not to be any music used, just the line of text you'd see on screen, and I decided because [the video] would be aired that I would try and clear copyright on it. And first of all it took a long time to find out who held the rights for it, even within Sony, like who actually was responsible for assigning permission. And that was with the help of a friend of mine, a feature film producer who put me on to lawyers who were very helpful. And then they decided that they wanted $2,000 U.S., which would have been a huge portion of the budget for this line of text that takes about two seconds to go across the screen. The total budget of my video was just over $14,000. So we went back and forth, back and forth, back and forth and eventually I think they charged $200 U.S.—that was just for a particular territory, and over a very limited period of time. So in the end I got a much better deal.

But it struck me how copyright really is meant to work from corporation to corporation. To do independent work, well, for one we don't have producers, generally, and two, the dividends are so small for the copyright holding companies like Sony, that they don't really want to deal with us. So it's easier for them to say no, you don't have permission than to go through this elaborate process, which then will give them something like $200 in the end.

— Richard Fung, video artist and faculty, Ontario College of Art and Design, Toronto, interview with Kirsty Robertson.

public archives alike may charge photo reproduction fees that are simply unaffordable to many filmmakers. Even though more footage from television or commercial cinema than ever is available via the Internet (and the National Film Board is to be commended for sharing more and more of its films online), permissions prices for reuse of clips are extremely expensive, well beyond the means of many beginning or independent filmmakers. And these clips are not mere decoration: as filmmaker Walter Forsberg says, "I wasn't around before 1980, so if I'm going to show those times, I have to use somebody else's footage."[8] Public institutions such as the NFB and the CBC could enhance their reputation by establishing and publicizing aggressive sliding scales for permissions, tied to the level of income generated by the project. They could even look into the Creative Commons licence as a way of distinguishing between commercial and non-commercial use. After all, that material was paid for by taxpayers in the first place. The CBC archive should be available to the public for non-commercial use and on a sliding scale for commercial uses. Of course, such projects have large technological and legal overheads for institutions strapped for money—but if those bodies started sharing their wealth, they would only increase the number of allies they have in the country.

12. VISUAL ARTS

The visual arts in Canada have a tradition of collective action and advocacy, and within the art community, copyright has generated a number of controversies. At the centre of discussion and action have been exhibition rights; moral rights; appropriation, parody, and satire; and resale rights. This chapter focuses on those issues, but more general information for artists can be found in Part II of this book, particularly in chapters 4, 5, and 8. Chapter 8 features a list of legal clinics offering advice to artists for free or for a nominal fee.

Exhibition Rights

One peculiarity of the art world is that many visual artists reach their most desired audiences and markets not through the reproduction of their work (as do writers and many musicians), but through the exhibition of original or limited edition work. So it makes sense that in 1988 Parliament gave visual artists the exclusive right to exhibit their work or authorize its exhibition; since the Status of the Artist Act came into force in 1992, artists have been able to collectively negotiate standard rates for exhibition in Canada.

When individual artists sell a work, either to a collector or to a gallery, they are understood to be selling only the rights in that physical object, unless a contract explicitly states that they are also transferring or licensing reproduction, exhibition, or moral rights. The exhibition right, set out in section 3(1)(g) of the Copyright Act, grants the artist the sole right "to present at a public exhibition, for a purpose other than sale or hire, an artistic work created after June 7, 1988, other than a map, chart or plan." Exhibition does not involve reproduction like a number of the enumerated section 3 rights, but we can think of it as an analogue to performance or communication to the public: just as writers and composers have a right

> Our argument essentially is that [galleries] have to adjust their priorities. They can't go to their staff and say, take half a wage to do the work you're doing. And the argument we're putting forward to them that they don't really like, is that we're peers and that we're an equal participant in that economy. Without us they wouldn't have their exhibitions, just as without a curator they wouldn't have somebody to organize the exhibitions. So we deserve to be paid on a level that's eventually on par with the senior people working in the system. You know, what curators get paid, and directors get paid, etc. . . . And the whole idea is that exhibitions enhance your sales, or enhance your academic career, or enhance your ability to get grants, and all that's true. But it still doesn't account for the fact that we're participating in an economy and we contribute to it.
>
> —Karl Beveridge, artist, CARFAC negotiator with the National Gallery,
> Toronto, interview with Kirsty Robertson.

to benefit from performances of their works, visual artists have an analogous right.

So the question is, How much is the exhibition right worth? Canadian Artists' Representation/Le Front des artistes canadiens (CARFAC) and Regroupement des artistes en arts visuels du Québec (RAAV) took the National Gallery of Canada to court for bargaining in bad faith when it refused to proceed with negotiations for binding minimum exhibition fees. A decision in the artists' favour was reversed by the Federal Court of Appeal in March 2013.[1] CARFAC and RAAV will no doubt continue to fight, but collective bargaining for artists' fees currently seems an unlikely prospect. It must be noted, furthermore, that the exhibition right is in practice only useful to artists who are well established. It does not apply to commercial galleries, or indeed to any show that includes works for sale. Thus it only kicks in for those whose work is included in curated shows at public galleries—a distant dream for emerging artists. Also on the topic of relations between galleries and artists, we might note that the Supreme Court decision in *SOCAN v. Bell* (2012) that iTunes music clips are fair dealing (see chapter 9) may undermine CARFAC and RAAV's claim that thumbnails on websites or in catalogues ought to trigger copyright payments. The balance of rights between galleries and artists is certainly favouring the galleries at present.

Moral Rights

Since 1988 some visual artists have had a slightly special status when it comes to moral rights. Whereas writers or filmmakers have to show damage to their honour or reputation to win a case of moral rights infringement, section 28.2(2) states, "In the case of a painting, sculpture or engraving, the prejudice referred to in subsection (1) shall be deemed to have occurred as a result of any distortion, mutilation or other modification of the work." The Act goes on to stipulate, in section 28.2(3): "(*a*) a change in the location of a work, the physical means by which a work is exposed or the physical structure containing a work, or (*b*) steps taken in good faith to restore or preserve the work shall not, by that act alone, constitute a distortion, mutilation or other modification of the work."

Even before this section was added to the Act, the 1982 case of *Snow v. Eaton Centre* in the Ontario High Court of Justice showed that courts were likely to recognize moral rights in unique visual works of art (see chapter 4). The Supreme Court set a higher bar for moral rights infringement in *Théberge v. La Galerie d'Art du Petit Champlain* (2002), which concerned the reuse of reproductions of paintings: the court was not convinced by Claude Théberge's claims that his reputation had been damaged by the actions of the gallery in question, given that he had already authorized them to make thousands of posters of his work.

In the *Théberge* case (also discussed in chapter 4), the artist had signed a contract with a gallery for the reproduction of his paintings on posters, postcards, and other stationery products. When he found that the gallery was selling his images on a canvas backing—to which they had been transferred from a poster by a new chemical process—he applied for an injunction and seizure of the offending objects. Ultimately, the Supreme Court found that there was no infringement in this case, because there was no reproduction.

> An expansive reading of the economic rights whereby substitution of one backing for another constitutes a new "reproduction" that infringes the copyright holder's rights . . . tilts the balance too far in favour of the copyright holder and insufficiently recognizes the proprietary rights of the appellants in the physical posters which they purchased. (para. 28)

Many artists find this case disturbing. The gallery made a new thing, which it sold for a higher price, presumably skimming off the extra profit; the court's idea that this did not constitute reproduction may seem mere sophistry. However, there was indeed no reproduction: the original poster paper was blank after the image had been transferred to the canvas. An analogous situation occurs when an artist uses pages from a book as the basis for a collage or an installation. If there is no reproduction, there is no infringement. It is a fundamental principle of the law. If the *Théberge* case had been decided the other way, practices such as collage or found-object art might have been in jeopardy. Artists might well see some merit in the court's statement that "once an authorized copy of a work is sold to a member of the public, it is generally for the purchaser, not the author, to determine what happens to it" (para. 31).

Appropriation, Parody, and Satire

Collage, a major genre of the twentieth century, is a pre-digital form: whereas a Robert Rauschenberg might glue together material bits and pieces (as he has said, "I think a painting is more like the real world if it's made out of the real world"), an artist in the twenty-first century is likely to make collage out of reproductions: printouts of material found online, or manipulated digital files. Such "appropriation" is a major aesthetic in today's art and culture. As more and more of us live in highly human-made environments, those are the worlds we seek to represent or see represented. As the Appropriation Art coalition stated in its argument for fair dealing rights for artists: "Aspects of [popular culture] are often reproduced as part of the work of art, but in such a way that the subject is transformed. . . . The new works that are produced comment on the world in which we live and reflect the nature of creativity itself."[2] Journalist Roberta Smith eloquently elaborates the social benefits of appropriation art:

> Our surroundings are so thoroughly saturated with images and logos, both still and moving, that forbidding artists to use them in their work is like barring 19th-century landscape painters from depicting trees on their canvases. Pop culture is our landscape. It is at times wonderful. Most of us would not want to live without it. But it is also insidious and aggressive. The stuff is all around us, and society benefits from multiple means of staving it off. We are entitled to have artists, as well as

political cartoonists, composers and writers, portray, parody and dis-
sect it.[3]

In the past, appropriation art has had a very precarious status in Canadian law. The only case on point, *Michelin v. CAW*, disqualified parody from
the fair dealing category of criticism (see chapter 5). However, in *CCH v.
Law Society of Upper Canada* (2004) the Supreme Court demolished the fun-
damentals of the *Michelin* logic. "Fair dealing," the highest court declared,
"must not be interpreted restrictively" because it is a way of recognizing
users' rights in the copyright balance. In *Alberta v. Access Copyright* (2012)
and *SOCAN v. Bell* (2012), the court affirmed the broad definition of fair
dealing purposes. The final nail in the coffin of *Michelin* is copyright reforms
in 2012 that added parody and satire to the list of fair dealing purposes.

How does a visual artist make use of fair dealing? In chapter 4 we explain
how to analyze whether a use of a copyright work is substantial, and in chapter 5 we explain how to go through a fair dealing test. The first part of the
fair dealing test is the purpose of the work, and it will be interesting to see
how courts will interpret the new categories of satire and parody. After all,
not all appropriation clearly fits those categories: some may be tribute, or
simply pastiche or collage.

As we speculate about how Canadian courts might interpret the new
fair dealing purposes, we might look at the various ways U.S. courts have
dealt with parody. In *Campbell v. Acuff-Rose Music* (1994), concerning 2 Live
Crew's rap version of Roy Orbison's "Oh Pretty Woman," the U.S. Supreme
Court was convinced that the 2 Live Crew version was a parody and thus fell
in the window of fair use. Interestingly, the successful legal strategy was
a bit of a stretch from an artistic point of view: many who heard the song
probably didn't know the Orbison original, so it could not have served as
a parody for them—nor was it clear that 2 Live Crew had parody in mind.
Similarly, when Alice Randall rewrote *Gone with the Wind* from the slaves'
point of view and called it *The Wind Done Gone*, she might have thought of it
as a critique or "writing back," but her lawyers called it a parody. They were
able to get a preliminary injunction lifted in 2001.[4]

If these two cases embrace a broad definition of parody, we find a different approach in *Rogers v. Koons* (1992). In that case, a New York district court ruled that Jeff Koons's sculptures based on Art Rogers's photographs of a couple with an armful of puppies constituted infringement. Koons argued the sculpture constituted parody of the "banality" of popular taste. The judge said that the sculptures constituted satire of social attitudes rather than parody targeting Rogers's work—in other words, the court decided that Rogers's work specifically was not required for Koons to make his point.[5] If such a case were heard in Canada today, artists could claim fair dealing under the "satire" umbrella. But they might still not be home-free: as *CCH* puts it, "If there is a non-copyrighted equivalent of the work that could have been used instead of the copyrighted work, this should be considered by the court" (para. 57).

Besides fair dealing, there is one other avenue for defending appropriation art: section 29.21, added to the Act in 2012, allows remix in any medium for non-commercial purposes. This exception will not help professional artists, but it may provide a safe space for students to experiment and learn. See the discussion in chapter 10.

Resale Rights

In sixty-seven other countries in the world, visual artists receive 5 per cent of the proceeds of resale of their work. CARFAC and RAAV have been working to bring resale rights to Canada, so far unsuccessfully. As they point out, in Canada this right would particularly benefit Inuit artists, whose work often escalates wildly in price from the first time they sell it.[6] Resale rights are conceptually analogous to royalties paid to musicians and writers. This right would bring visual artists more in line with the way other creators are paid: the more the market values their work, the more they would earn.

13. CRAFT AND DESIGN

The term "design" can be applied to many different practices—from graphic design to designs of mass-produced objects to (our focus here) the design of useful objects such as shawls, cups, or chairs produced in small numbers by artisans.

Graphic design—for example, the design of book covers, signage, and advertisements—is basically commercial visual art, which copyright law treats under the rubric of visual art (see chapter 12). Much graphic design is done by employees, in which case the work belongs to the employer unless otherwise stipulated; other graphic design is done by independent contractors or freelancers. The major copyright issue in the latter case is to have a clear contract between the parties that states what uses exactly are being paid for. This wording does not have to be in legalese; a simple invoice stating permitted uses is sufficient.

Industrial design—that is, design that culminates in mass-produced objects—is generally not automatically protected as intellectual property, and does not fall under the Copyright Act. But it can be registered under the Industrial Design Act with the Canadian Intellectual Property Office and thus protected from copying for a term of ten years.[1]

The Canadian Copyright Act does provide specific treatment for some categories of design. Section 64(1) defines "design" as "features of shape, configuration, pattern or ornament and any combination of those features that, in a finished article, appeal to and are judged solely by the eye." If the object has a "function other than merely serving as a substrate or carrier for artistic or literary matter," the Act states, its design is copyrighted if no more than fifty such objects are made. Hence a saddle, bowl, or dress is only covered by copyright if made in very low quantities. The Copyright Act thus implicitly defines industrial-scale production as production in quantities larger than the cut-off of fifty. Presumably the rationale is that there is a public interest in rapid innovation in industrial design, which can be facilitated by relatively weak intellectual property protection, whereas designs of useful objects outside the mass market (or prior to entering the mass market) ought to have some protection.

In all cases, drawings or sketches of the designs are covered by copyright as such: you might be able to imitate a mass-produced purse without permission, for example, but you can't copy its designer's drawings. The Copyright Act, under section 64(3), also includes exceptions to the fifty-object limit. For example, architectural works, patterned fabric, and graphics applied to the face of articles (such as images on T-shirts or mugs) retain copyright protection even if they are mass-produced.

Note that U.S. and Canadian law noticeably differ in the area of design.[2] The U.S. Copyright Act states, "The design of a useful article . . . shall be considered a pictorial, graphic, or sculptural work only if, and only to the extent that, such design incorporates pictorial, graphic, or sculptural features that can be identified separately from, and are capable of existing independently of, the utilitarian aspects of the article."[3] This "separability requirement" has been understood to mean, for example, that clothing design is not covered by copyright. That is, the design features are not simply attached to an underlying utilitarian article; the article embodies the design.[4] Canadian law has no such requirement; if a clothing designer in Canada makes, or authorizes the manufacture of, articles in numbers fewer than fifty-one, the work is copyrighted, whether or not the design is "separable" from the function of the article.

Are the law's design protections of use to artisans? In practical terms, rarely. Few artisans can afford to litigate the copying of their designs. Industrial

I design hats and make more than fifty of each kind. Can I use the law to protect myself from imitators?

You could obtain protection through the Industrial Design Act. The registration process is fairly simple (see the description at the Canadian Intellectual Property Office: www.cipo.ic.gc.ca). After an application is made and a fee is paid, the design goes through an examination process. If granted, the registration gives the owner exclusive rights for a period of ten years. Under section 11(1) of the Industrial Design Act, the owner has the exclusive right to

(a) make, import for the purpose of trade or business, or sell, rent, or offer or expose for sale or rent, any article in respect of which the design is registered and to which the design or a design not differing substantially therefrom has been applied; or

(b) do, in relation to a kit, anything specified in paragraph (a) that would constitute an infringement if done in relation to an article assembled from the kit.

However, it appears that few artisans make use of this tool; for one thing, it would be onerous to register a range of variations of designs. A more common strategy would be to unofficially brand the product through promotion, and to keep coming up with new designs to stay ahead of the imitators.

design registration is onerous and impractical for a truly innovative independent designer who has many different objects in production. Most artisans find it difficult if not impossible to make a living by keeping their quantities below fifty-one (a good candlestick or belt design, for example, will be replicated for many craft shows and retail outlets), so copyright is only helpful as their work approaches one-of-a-kind status. The legal distinction between a potter and a sculptor, or between a "useful" and a "useless" object, may seem to be rather arbitrary and unfair: a printmaker producing an edition of five hundred prints holds copyright in the work, but a maker of fifty-one identical art vases may not, simply because the vases can hold flowers.

This awkward fit between law and artisanal circumstance has resulted in a culture of what we might call "folk copyright" in crafting circles. There is a widespread fear of rip-offs among artisans and designers.[5] We are living in an era in which North American consumers can buy beautifully designed articles for their homes for very little money—articles produced in countries with lower labour costs than ours. The design of many of these inexpensive products is inspired or even copied, without authorization, from folkloric styles or the work of individual designers. In addition, trends in North American–produced crafts often derive from one artisan's hit design.

In this context, there is an emerging feeling in craft and design communities that something needs to be done. Crafting websites are loud with assertions about copyright. But what gets called copyright in these discussions is often rather murky. For example, quilt designer Kathleen Bissett convinced the Central Canada Exhibition that quilts entered in competitions without the permission of the designer infringed copyright. This is not the case: while the particular wording or drafting of a pattern is copyrighted, the steps and process are not, and thus according to the law, designers have no say in the use or embodiment of their patterns. (They do have a moral right to attribution.) And yet the claim, like other similar assertions untested in court, is becoming a new normal in crafting circles.[6]

On the other hand, other craftspeople are ambivalent about wielding the law because they envision their artistic practice in terms of sharing and

Jewellery-makers Wade Papin and Danielle Wilmore run Pyrrha Design Inc. in Vancouver. When a Saskatchewan accessories retailer ordered jewellery based on their designs to be made in Taiwan and started selling the products in Canada for a much lower price, Pyrrha Design took him to court. The legal question was whether jewellery was art or design: if design, it would not be protected by copyright, because Pyrrha produced the pieces of jewellery in quantities larger than fifty. Pyrrha's lawyer accordingly argued that the jewellery was useless, mere adornment: in other words, art.

It took five years and tens of thousands of dollars, but eventually, as the case slowly moved towards trial, the imitator agreed to settle. While Papin and Wilmore won satisfaction and compensation, the settlement meant that the case never went to court. It remains unclear whether jewellery, in Canadian law, constitutes art or design.

Source: Alexandra Gill, "It Glitters. It Sparkles. It's Art," *Globe and Mail*, 5 April 2006.

giving, not market exchange. They know that they base their work on established design vocabularies and traditions, used without permission. Potter Jane Thelwell declares, "I make thin mugs. That's my calling card." But she also acknowledges that "I'm not the first person on the planet to throw thin, there's me and about a billion Chinese guys who are now dead who used to do it."[7] While she seeks payment for her pottery, and credit from others for techniques they learned from her, she does not expect a copyright relationship to her work. Many participants in the crafting renaissance, amateurs and small commercial producers alike, are actively hostile to the whole idea of copyright. As art scholar Kirsty Robertson puts it:

> The debates over copyright within crafting communities are particularly thorny, jumping as they do from notions of a common shared history that should be open and welcoming to all, passing through the idea that as a gendered pastime crafting is regularly devalued—

When you're trying to learn a technique, you learn to replicate something that you're seeing. All of the projects in school, that's what they are. But you're encouraged very early on to put your own stamp on it: we all know about not wanting to be copycats. You very quickly internalize this fear of being perceived as copying, even though copying is how we learn. It's very paradoxical.

And the other thing is, we're all working with the same tools. We're all working with the same colour palette, we're all working with the same history of techniques, and we're all working three-dimensionally in the round: the material lends itself to certain things. So there are things, if you know glass, you can believe that people would come up with independently, because it's just what glass wants to do. Christmas balls are a very good example: everyone makes a blown globe of some sort, because glass just wants to be a bubble . . . and that's the fundamental thing, the bubble. You can't claim that somebody's copying, because everybody's making a bubble from glass. You can't know who did it first.

— Susan Belyea, glassmaker, interview with Laura Murray, Kingston, Ontario

something its practitioners should work against—to more recent arguments that there are lucrative opportunities for professional crafters and designers that need to be protected through the copyrighting, patenting and trademarking of designs and processes.[8]

The place of the law in artisanal environments is in a contested state not because the law has changed, but because the Internet allows new pathways for copying, sales, and debate, because low-cost Asian manufacturing has blurred the very definition of "craft," and because in this context values are being tested and contested.

Is a sewing pattern covered by copyright?

As an image, yes. You can't make a wallpaper pattern from it without first getting permission.

As a process, no. When you make a dress from that pattern, you are essentially following its instructions; you are not copying the pattern itself. The Copyright Act, section 64.1(1)(d), reminds us that copyright does not protect "any method or principle of manufacture or construction." By the same logic, we can all rejoice that making a cake is not an infringement of copyright in the recipe—even if the baker sells the cake.

If the pattern bears a notice saying that you cannot use it to make products for sale, this notice may be considered a contract to which you agree by buying the pattern or using it. But whether the stricture is enforceable is not clear.

14. JOURNALISM

Journalists face copyright issues coming from two directions: they need to assert their own rights in a media marketplace increasingly dominated by large corporations, and they need to have strong fair dealing rights so they can tell the stories they uncover. The public interest is served when journalists can support themselves well enough to develop their expertise and skills over a lifetime. But it is also served when the citizenry has broad access to the news record: news is a powerful democratic tool and is the embodiment of shared history. The Copyright Act and the courts are quite sensitive to the various rights and interests at play in this profession, but concentration of media ownership poses serious challenges to a productive balance of interests.

Who Owns What?

Who owns copyright in a given newspaper column or magazine story? Not surprisingly, the answer is, it depends. There are clear default rules for both staff writers and freelancers, but contracts can and often do trump those. Section 13(3) of the Copyright Act applies to staff writers along with bank workers, civil servants, and all employees: it provides that, without any agreement to the contrary, the employer is the first owner of work done in the course of employment. Employees cannot benefit from or control reuse of material they have created in the course of employment.[1] Staff journalists do have one special right compared to other employees: "in the absence of any agreement to the contrary," they have the "right to restrain publication" beyond newspaper or magazine form. It is not clear what this right is worth, but it might be something that staff journalists should try not to sign away.[2]

Freelancers are not employees, and hence they are the first owners of their work.[3] They are governed by the general rule on copyright assignments in section 13.4, which states: "The owner of the copyright . . . may assign the right, either wholly or partially, and either generally or subject to limitations relating to territory, medium or sector of the market or other limitations . . . and either for the whole term of the copyright or for any other part thereof." The journalistic convention, until recent years, was for freelancers to license "first North American rights" to the first buyer: that is, you might be paid one dollar per word in exchange for first publication rights in a New Brunswick newspaper. You could then go on to sell reprint rights in British Columbia, translation rights in Germany, and broadcast rights in Japan. However, with the rise of media conglomerates and digital technologies, freelancers have lost many opportunities to redeploy their work. While new possibilities exist for syndication or selling digital rights, freelancers face constant strong-arming. For example, Canwest/Global introduced contract terms in 2004 transferring all rights "throughout the universe, in perpetuity"; the individual who refuses to sign such a contract risks losing the job to someone who will.[4]

According to a study by the Professional Writers Association of Canada (PWAC), the average freelance writer lost 26 per cent in purchasing power in the decade from 1995 to 2005.[5] No doubt the trend has persisted. This problem is not directly caused by copyright, nor can it be resolved by copyright alone. But it is all the more important for freelancers to know that they have more rights than media outlets often let on. The increasingly standard publisher demand for an exclusive grant of all rights represents a unilateral change to traditional industry practice. It is backed by bargaining power, not by anything in the Copyright Act.

Despite the media oligopoly, PWAC maintains, "There is no such thing as a non-negotiable contract, so negotiate."[6] The association's main overall advice is to make sure that all licensed rights are enumerated in writing and separately paid for. Following this advice means that all rights not explicitly mentioned in a contract are reserved to the writer. Thus, "selling a story" to *Canadian Gardening* might actually mean licensing a one-time print publication, website display for a set time span, and resale to commercial archival databases on specified terms. The writer could then go on to sell other rights, such as translation or secondary publication, to other publishers. This distinction between an exclusive right and a non-exclusive right is important: it is to an author's advantage to enter into non-exclusive arrangements, or to make very sure indeed that the price for an exclusive right is high enough to make it worthwhile. If the magazine wants exclusive web publication rights, for example, it should pay more than if it is allowing the writer to license

Many young journalists leave the profession. One of them says: "Every year the rates get a little worse, and every year the real features give up a little more space to packaged stuff. Suffering for art's sake is one thing, but suffering for charts about $5,000 barbecues? Life's too short."

Source: Posting to Canadian Magazines discussion, by anonymous journalist, 29 May 2006.

Robertson v. Thomson (2006)

The situation:

In 1995 Heather Robertson wrote two articles for the *Globe and Mail*. Subsequently, the *Globe* included them without her permission in three electronic databases, and Robertson sued for copyright infringement. The *Globe* argued that it had copyright in the newspaper as a compilation, and thus had the right to this reproduction.

The result:

In a 5–4 decision, the Supreme Court dismissed the *Globe*'s claim:

> We . . . agree with the Publishers that their right to reproduce a substantial part of the newspaper includes the right to reproduce the newspaper without advertisements, graphs and charts, or in a different layout and using different fonts. But it does not follow that the articles of the newspaper can be decontextualized to the point that they are no longer presented in a manner that maintains their intimate connection with the rest of that newspaper. . . . These products are more akin to databases of individual articles rather than reproductions of the *Globe*. Thus, in our view, the originality of the freelance articles is reproduced; the originality of the newspapers is not.

The court said that the *Globe* could reproduce aggregated pages of its newspaper if the emphasis were on the compilation as a whole, but unauthorized reproduction of articles by freelancers in a search engine enabling more fine-grained and independent searching by individual article would be infringement.

web publication rights to others as well. Similarly, prices for permanent rights transfers and for temporary licences might differ considerably. *You get what you pay for* should be the freelancer's motto.

The *Robertson v. Thomson* case of 2006 confirmed that a publisher holds only those rights that are explicitly itemized in the contract. Absent such a specification, all remaining rights are reserved to the freelancer. Philosophically, however, the issue is much more complex, relating to the nuanced layering of rights. The *Globe and Mail* exercised considerable skill and judgment in its editorial, selection, and arrangement functions, and does indeed hold copyright in the newspaper as a whole. In a dissenting opinion, Madam Justice Rosalie Abella argued that as new media allow new structuring of content, the newspaper's copyright should not be eroded: "The conversion of a work from one medium to another will necessarily involve changes in the work's visual appearance, but these visual manifestations do not change the content of the right." The puzzle is that arguments on both sides in this case were based on the idea of technological neutrality: the freelancers wanted to be paid for electronic articles, and the newspapers wanted to utilize new modes of archiving and searching.

Furthermore, the dissent in this case made it clear that a third interest is at play in the battle between publishers and freelancers: the public interest. As Abella put it, "The public interest is particularly significant in the context of archived newspapers. These materials are a primary resource for teachers, students, writers, reporters, and researchers. It is this interest that hangs in the balance between the competing rights of the two groups of creators in this case, the authors and the publishers."[7] Given that the larger publishing enterprises have for some years been pushing their contributors into contracts that permit archiving, this problem may be largely retrospective, but that does not make it less grave: the *Robertson* decision could be interpreted such that large swaths of Canadian journalism from the 1940s through the 1990s will be inaccessible to all but specialist researchers just at a time when digital technologies could give the pieces broader exposure. We hope that legal and practical solutions are found to enable digitization of journalistic heritage, including literary journalism and small-press publications.[8]

There are other ways that publishers constrain the options of users. At the same time as a large media corporation might be demanding all rights from its content providers, it is licensing fewer and fewer rights at the other end. This is not an area where claims about consumer piracy can gain any traction: it is so clear that corporate piracy is the main problem. Thus someone who pays for access to the *Toronto Star*'s Pages of the Past archive, which goes back to 1892, must agree to ignore that a great deal of the material to be found there is in the public domain: "Distribution, transmission, or republication of any material from www.thestar.com is strictly prohibited

In June 2007, Gordon Murray and Carel Moiseiwitsch, struck by what they saw as a consistent pro-Israel, anti-Palestine bias in coverage in the *Vancouver Sun*, manufactured twelve thousand copies of a parody issue of the paper. With other activists, they handed papers out to morning commuters. The main headline was "Celebrating 40 Years of Civilizing the West Bank," accompanied by a photo of a man in a headscarf facing a massive bulldozer. The photo caption read, "Israeli military bulldozer brings the many gifts of civilization to a primitive resident of Nablus in the West Bank."

Canwest Mediaworks Inc., then the owner of the *Sun*, was not amused. It called the paper "a cowardly act of public deception," alleging trademark and copyright infringement. Murray and Moiseiwitsch refer to the legal action as a SLAPP, strategic lawsuit against public participation, as it surely was. After all, it's not that people who read more than a handful of words would seriously have mistaken the parody for the *Sun* itself. This was a mobilization of copyright to shut critics up.

The case sputtered out when Canwest went bankrupt; its new owner has not reactivated the suit. And it won't: the new inclusion of parody as a fair dealing purpose in the Act now directly protects such uses.

Sources: Laura Murray and Craig Berggold, "See You in Court: Can Canadians Practice Parody?" *FUSE Magazine* 32.2, March 2009; http://seriouslyfreespeech.ca.

without the prior written permission of Toronto Star Newspapers Limited."[9] Beyond the public domain nature of older material, educational uses of individual articles would be fair dealing, so this claim is fallacious in more than one way. The British Library noted in 2006 that "twenty eight out of thirty licences [for electronic resources] offered to the British Library and selected randomly were found to be more restrictive than rights that currently exist within copyright law."[10] Large media corporations are amassing more and more rights by fiat at the expense of writers and readers alike.

Users' Rights

Negotiating over ownership and owners' rights appears to be the most prominent copyright issue facing journalism today. Users' rights may be noticed most as an issue when exaggerated owners' rights claims are made, as we saw above. But as an institution, journalism has long been dependent on the unauthorized reproduction of facts, ideas, and even substantial portions of copyright material. Some of this recycling simply falls outside of copyright law—story ideas, for example, spin from one media outlet to another in breathless flurries and then die out. But other practices present exceptions to an owners' rights logic. By tradition of the trade, media outlets often lift substantial portions of copyright material from each other or from the entertainment industry: movie stills, images from websites, and magazine covers, for example. That they don't ask for permission in these situations is not an oversight. Many of these excerpts essentially serve as free advertising for their source: Janet Jackson almost nude on the cover of a magazine is not only free salacious content for the somewhat staid *Globe and Mail*, but also promotion for both Jackson's album and the magazine.[11] But more serious democratic principles are also at stake. When the *Winnipeg Free Press* covers an incident in an election campaign, it does not ask permission to quote candidates or members of the public, or to reproduce the text of a ten-year-old letter showing a candidate up to be a liar. If the CBC then picks up the *Free Press*'s scoop, it does not ask permission to reprint or pay for the *Free Press*'s labours, although by convention it cites the source.

Glenn Gould Estate v. Stoddart (1998)

The situation:

In 1956 reporter Jock Carroll interviewed and photographed pianist Glenn Gould for an article for *Weekend* magazine. In 1995, after Gould's death, Carroll published a book with Stoddart featuring a wider sampling of photographs and interview materials from those sessions. The Gould estate sued.

The result:

The Appeal Court found that Carroll owned the copyright in the photographs, and also in the interviews—since Gould's unscripted words did not meet the fixation requirement of copyright, and since there was no evidence of any special contractual arrangements between Gould and Carroll. Carroll was the sole author of these materials, and as such held sole right to publish them. "Once Gould consented, without restriction, to be the subject matter of a journalistic piece," the court held, "he cannot assert any proprietary interest in the final product nor can he complain about any further reproduction of the photographs nor limit the author of the journalistic piece from writing further about him."

The implications:

Nowadays the subject of a major interview project would probably demand a contract, and things would be clearer. But the Gould case states that absent a contract, unscripted words are not copyrightable, whereas their transcription is. This result is good for journalists with drawers full of old interview notes, but it shows a legal bias in favour of the written word that is more than a little problematic. Copyright law does, however, protect more considered spoken words: storytellers could almost certainly argue successfully in court that their words constituted "performers' performances," protected under section 26. And an interview subject whose words were used in unexpected ways might have recourse beyond copyright via privacy or implied contract arguments.

The principle here is that it is the media's job to reveal the truth to their listeners or readers, and to do this they need to be fast and unhampered. The established journalistic custom of borrowing is formalized as fair dealing in the Copyright Act (with the specification that the source must be mentioned). Case law suggests that journalists should not be hesitant to practise fair dealing. In *CCH v. Law Society of Upper Canada* the Supreme Court endorsed the consideration of the custom or practice in a particular trade or industry as one of the fair dealing criteria. If the borrowing in question follows normal newsroom practice, then it would most likely fall within the provision of fair dealing (depending on the other circumstances of its use). *Allen v. Toronto Star* indicates that journalistic fair dealing includes very substantial uses, if they are necessary to the reporting end.

Allen v. Toronto Star (1997)

The situation:

In 1985 Liberal Party Member of Parliament Sheila Copps made the cover of *Saturday Night* magazine, dressed in black leather and sitting astride a Harley. In 1990 the *Toronto Star* ran an article discussing Copps's change of image as she ran for leadership of the Liberal Party—illustrating this article with a reproduction of the *Saturday Night* cover, contrasted with a more recent photograph of Copps dressed and posed more sedately. *Saturday Night* did not object to the unauthorized reproduction of its cover, but Jim Allen, the photographer whose work was featured on the cover, sued the *Star* for copyright infringement.

The result:

Allen won at trial, but the *Star* appealed. The higher court noted that Allen did not own the copyright on the cover in question, which is what the *Star* had reproduced—even though he still owned copyright in his photograph. Furthermore, it held that the *Star*'s use was fair dealing for purposes of news reporting. "In our view," the court stated, "the test of fair dealing is essentially purposive. It is not simply a mechanical test of measurement of the extent of copying involved." In this case, an entire magazine cover was held to be fair dealing, given "the nature and purpose of the use": it was central to the topic of the story, and it did not give the *Star* an "unfair commercial or competitive advantage over Allen or *Saturday Night*."

15. EDUCATION

Educational institutions are major incubators for future creativity and economic growth. They cultivate both the ability and urge to seek out works of human expression, and the ability and urge to create and critique them. Considering copyright's roots in the 1710 British Act "for the Encouragement of Learning," we should expect copyright to facilitate education. Furthermore, the education sector is a major market for copyright products. Schools and post-secondary students buy hundreds of millions of dollars' worth of books, software, and other copyright material every year. For example, in 2011–12 the Queen's University Campus Bookstore sold $7 million in new textbooks—holding fairly steady from five years earlier—and another $1.5 million in licensed coursepacks, used books, rentals, and digital materials.[1] In 2012–13, the Queen's University Library allocated $1,173,000 for books, $455,000 for print periodicals and journals, and $6,427,392 for electronic resource purchases and subscriptions.[2]

Given the scale of the educational market, publishers are anxious about any erosion of it. Leery of photocopying practices, and now equally anxious about the Internet distribution of teaching materials, they have advocated narrow understandings of fair dealing and expanded use of collective licensing. Meanwhile, though they might have asserted the users' rights articulated by the Supreme Court in *CCH v. Law Society of Upper Canada* (2004), school boards and post-secondary institutions have remained overly concerned about liability and continued to pay hefty licence fees, thereby imposing costs on taxpayers or students that are not justified. As we will emphasize in this chapter, both legislative reform and judicial decisions promise to improve this situation, assuming that they are translated into change at the level of the local institution.

Beyond the use of purchased physical books and DVDs, so far still the bread and butter of routine educational materials, teachers and students have often confused three distinct mechanisms that allow them to make uses of other copyright materials without asking and without infringement. The first mechanism is licences that institutions have with individual publishers or collectives such as Access Copyright, which grant certain copying rights in exchange for fees. Authorization for use has already been given in this case, but users may or may not be aware of that; they mainly know that it's okay to use the material. The second mechanism is specific educational exceptions in the Copyright Act; the third is fair dealing. These last two are provisions in the Act that make authorization unnecessary.

In the wake of Bill C-11 and the Supreme Court cases *Alberta v. Access Copyright* and *SOCAN v. Bell* in 2012, fair dealing has acquired a greater weight than ever before, so we will discuss it first. Then we will consider specific exceptions and licensing. First, though, we want to note that a major change of Bill C-11 that positively affects educational use of copyright material is the reduction of statutory damages awards. Not only have the users' rights provisions become strengthened, but the penalties should a use not qualify have been substantially reduced. This changes the whole risk calculus for educational institutions.[3]

Fair Dealing

As we have seen, *CCH v. Law Society of Upper Canada* (2004) explicitly held that users' rights are a part of Canadian copyright law. Since the case concerned library photocopying, the ruling specifically extended fair dealing concepts to libraries. *CCH* made it clear that fair dealing can be practised in libraries, by library patrons offsite, by library workers on behalf of patrons, and even by lawyers and their employees in private, for-profit practice. Furthermore, in *CCH* the Supreme Court asserted that fair dealing existed independently of specific exceptions:

> As an integral part of the scheme of copyright law, the s. 29 fair dealing exception is always available. Simply put, a library can always attempt to prove that its dealings with a copyrighted work are fair under s. 29 of the *Copyright Act*. It is only if a library were unable to make out the fair dealing exception under s. 29 that it would need to turn to s. 30.2 of the *Copyright Act* to prove that it qualified for the library exemption. (para. 49)

It would follow from *CCH*, then, that an educational institution can rely on fair dealing in addition to the special educational exceptions. As described in chapter 5, *CCH* then went on to articulate a set of tests for fairness that can be applied in any setting.

A significant change to the Copyright Act in 2012 was the addition of "education" to the list of fair dealing categories. Section 29 of the Act now reads, "Fair dealing for the purpose of research, private study, education, parody or satire does not infringe copyright." While this amendment fell short of adding the more inclusive "such as" language that we proposed in our first edition, and while it did not incorporate the text of the fair dealing factors endorsed by the *CCH* decision, it should clear up a lot of the uncertainty that has plagued the implementation of fair dealing practices in educational institutions even after *CCH*.

The addition of education to the enumerated categories in section 29 is significant for a number of reasons. First, all of the activities that take place as part of the educational purpose of a school, college, or university are now likely within the ambit of fair dealing. This does not mean that any and all copying will satisfy the tests of fairness as laid out in *CCH*. But it does mean that the first step of fair dealing analysis will routinely be satisfied in the educational setting, and attention can now turn to the development of best practices standards that will satisfy the subsequent criteria. Second, the addition should add an increased level of comfort for many risk-averse educational administrators. Finally, broadening the fair dealing categories brings Canadian fair dealing and American fair use into much closer alignment. This will make it more feasible to look across the border for guidance and practical advice from U.S. educational institutions, which have historically taken much more proactive approaches to users' rights in the educational setting than their Canadian counterparts.[4]

As we have discussed earlier (see chapter 5), the other important 2012 development in fair dealing is the Supreme Court cases *Alberta v. Access Copyright* and *SOCAN v. Bell*. These cases build on *CCH* and lay to rest any lingering doubts that fair dealing is some sort of special exemption: whereas *CCH* dealt with the fairly rarefied context of a professional library for lawyers, these cases explored fair dealing in K–12 education and commercial digital music. Both of the cases assert (or rather reassert, following *CCH*) that the fair dealing purposes must be given a very broad definition. While the addition of education to the fair dealing purposes in the Act will reduce educational institutions' reliance on the categories of research and private study (the purposes considered in these two cases), the resounding language of the court on the breadth of fair dealing categories will echo in any future consideration of fair dealing in education.

In *Alberta*, Access Copyright had argued that educational use could not be research or private study if it was done at the behest of the teacher. But Justice Abella rejected that claim:

It seems to me to be axiomatic that most students lack the expertise to find or request the materials required for their own research and private study, and rely on the guidance of their teachers. They study what they are told to study, and the teacher's purpose in providing copies is to enable the students to have the material they need for the purpose of studying. The teacher/copier therefore shares a symbiotic purpose with the student/user who is engaging in research or private study. (para 23)

And she goes on:

. . . the word "private" in "private study" should not be understood as requiring users to view copyrighted works in splendid isolation. Studying and learning are essentially personal endeavours, whether they are engaged in with others or in solitude. By focusing on the geography of classroom instruction rather than on the *concept* of studying, the [Copyright] Board . . . artificially separated the teachers' instruction from the students' studying. (para 25)

Going forward, this holistic understanding of teaching and learning should have significant implications for the development of educational fair dealing policies. Instructors and librarians have always understood the intertwined relationships between teaching, learning, and educational materials, and these insights need to be better reflected in institutional copyright policies.

In practical terms, *Alberta* makes multiple copies for classroom use into a possible fair dealing—subject to the tests for fairness. "Teachers do not make multiple copies of the class set for their own use, they make them for the use of the *students*," it asserts (para. 29). So from each individual student's point of view, no more than one copy is made. The court furthermore insists that in assessing the "amount" factor, the proper inquiry is not "based on aggregate use, it is an examination of the proportion between the excerpted copy and the entire work, not the overall quantity of what is disseminated" (para. 29). That is, the analysis of the use of a portion of a textbook is based on the relation of the portion to *that particular book*, not on the whole realm of copying in an institution. Similarly, the *SOCAN* case determined that listening to music previews online on a service such as iTunes constituted research, focusing not on the more expansive use made by the commercial service but on the local uses made by potential consumers. Tests for the character and amount of the dealing are to be applied with respect to individual users and individual works.

Another significant finding in *Alberta* was that the Copyright Board's approach to the "alternatives to the dealing" fairness test was unreasonable. The Board had found that the schools had an alternative to the copying: they could have purchased more books. But the court found that "buying books for each student was not a realistic alternative to teachers copying short excerpts to supplement student textbooks" and that making copies of short excepts was "reasonably necessary to achieve the ultimate purpose of the students' research and private study" (para. 32).

As to the final criterion, "effect of the dealing on the work," Access Copyright claimed this factor went against fair dealing because of the decrease of textbook sales over the last twenty years. The Copyright Board had agreed with this approach, even though there was no evidence presented linking the decline in sales to the copying done by the teachers. In reversing, the Supreme Court observed that "other than the bald fact of a decline in sales over 20 years, there is no evidence from Access Copyright demonstrating any link between photocopying short excerpts and the decline in textbook sales."[5]

All in all, then, fair dealing in education has been massively bolstered. The court has shown no reservations whatsoever: it is now up to educational institutions to implement its decisions with energy and vision.

Educational Exceptions

The 1997 amendments to the Copyright Act contained a number of provisions applicable to educational institutions; several of these provisions were revised in 2012, and a couple of new ones were added.[6] These sections of the Act apply only to certain educational institutions: the Act's definition of "educational institution" includes publicly administered schools, community colleges, and universities, as well as anyone acting under the authority of these institutions, but does not include private schools or training institutes unless they are formally registered as non-profit.[7]

The exceptions define a particular set of circumstances in which infringement is excused, and they are generally offset by counter-limitations, record-keeping requirements, or other constraints. For example, a teacher can have students perform a play at a school assembly, but cannot have them perform it at a nursing home.[8] Thus, while some educational exceptions have their utility, and while they have historically been preferred to fair dealing by school boards and post-secondary administrators, these exceptions present many problems as a general approach to users' rights in the classroom. Their very multiplicity and complexity is confusing and disempowering for both students and teachers. This is why the *Alberta* ruling is so important: it recognizes that the classroom environment needs flexibility, and hence gives ample space to fair dealing above and beyond the particular exceptions.

The educational exceptions are enumerated in Table 12, but we will take extra time here to go over the new or updated ones.

Classroom Displays

Section 29.4, as previously drafted, had provided an exception for classroom displays on "a dry-erase board, flip chart or other similar surface" or on "an overhead projector or similar device." Under the revised Act, the display exception is cast in more technologically neutral terms. Section 29.4(1) now simply provides: "It is not an infringement of copyright for an educational institution or a person acting under its authority for the purposes of education or training on its premises to reproduce a work, or do any other necessary act, in order to display it." While the more purposeful language is a positive step, its benefits are negated where the work is "commercially available."[9]

Table 12. Exceptions Applicable to Educational Institutions

Common Name	What It Allows	Limitations	Section of Copyright Act
classroom displays	reproduction of a work for purposes of display	• work must not be commercially available in appropriate medium • must be for education "on premises"	29.4(1)
reproduction for exams	reproduction and telecommunication for the purpose of tests and exams	• work must not be commercially available in appropriate medium	29.4(2)
live performance	performing a work	• audience must be primarily students and educators	29.5(a)
performance of recordings and broadcasts	playing sound or film recordings or broadcasts	• audience must be primarily students and educators • copy must not be infringing or person responsible has no grounds to believe it is	29.5(b)(c)(d)
recorded news broadcasts	copying a news program for classroom use	• audience must be primarily students and educators • does not include documentaries	29.6
other recorded broadcast	copying broadcasts for classroom use	• must be destroyed or paid for within 30 days	29.7

Table 12. Exceptions Applicable to Educational Institutions (continued)

Common Name	What It Allows	Limitations	Section of Copyright Act
telecommunication of lessons	distance learning including copyright materials	• lesson must be destroyed within 30 days of course completion • educational institution must take measures to ensure limited distribution	30.01
digital copying	making digital copies under a licence for photocopies	• institution must have licence with a reprography collective • must take measures to ensure limited distribution • subject to later possible tariffs	30.02, 30.03
publicly available material (PAM) on Internet	make ordinary use of the Internet including public performance, reproduction, communication	• source must be named • not available if TPM or "clearly visible notice" present	30.04

Public Performance

Section 29.5 allows certain public performances "on the premises of an educational institution for educational or training purposes and not for profit, before an audience consisting primarily of students of the educational institution, instructors acting under the authority of the educational institution or any person who is directly responsible for setting a curriculum for the educational institution." The exception applies to the live performance of a work done primarily by students on the premises of the educational institution. Before the 2012 amendments, the exception applied to sound recordings and performers' performances embodied in sound recordings, and to broadcasts at the time of transmission, but not to films. Bill C-11 adds films, "as long as the work is not an infringing copy or the person responsible for the performance has no reasonable grounds to believe that it is an infringing copy." This change should help make it possible for teachers to incorporate all types of media into classroom instruction.[10]

Recorded News Broadcasts

Section 29.6 permits making a single copy of a news program or news commentary program (but not a documentary) for subsequent classroom use. Before the 2012 amendments, the exception only allowed keeping the copy for a year, at which time royalties had to be paid or the copy had to be destroyed. Bill C-11 removed the pay-or-destroy requirement. Other broadcasts (including documentaries) are covered by section 29.7, which retains the pay-or-destroy requirement after thirty days.

Telecommunication of Lessons

Section 30.01 provides a new exception for certain uses of a lesson. It begins with a threshold definition of "lesson" that is circular and confusing.[11] It goes on to provide that it is not an infringement of copyright

(*a*) to communicate a lesson to the public by telecommunication for educational or training purposes, if that public consists only of students who are enrolled in a course of which the lesson forms a part or of other persons acting under the authority of the educational institution;

(*b*) to make a fixation of the lesson for the purpose of the act referred to in paragraph (*a*); or (*c*) to do any other act that is necessary for the purpose of the acts referred to in paragraphs (*a*) and (*b*).

If the intention of this exception was to permit activities associated with technology-enhanced learning, it might have been simpler to just expand the definition of "premises." Whatever "lesson" does mean, the usefulness of this exception is negated by its limitations. First, under section 30.03(6)(a), the institution must "destroy any fixation of the lesson within 30 days after the day on which the students who are enrolled in the course to which the lesson relates have received their final course evaluations." Few institutions or teachers have the time or money to go to the trouble of preparing a lesson only to destroy it after the end of the term. It is also unclear how this provision would be enforced.

Furthermore, under section 30.01(6)(b), the institution is under an obligation to "take measures to limit the communication by telecommunication of the lesson to the enrolled students and other authorized persons" as well as, under section 30.01(6)(c), to take measures to prevent students from "fixing, reproducing or communicating the lesson other than as they may do under this section." What does it mean to take such measures, and how is this obligation to be enforced? We think that telling students they must destroy their course materials is simply not an acceptable practice; educators usually hope that students will refer back to materials from courses they have taken.[12]

There is additional uncertainty here because further regulations could set out even more requirements associated with this exception. It could have some application to streaming practices, but in general, given the expansion of fair dealing to include education, we would recommend that this section not be relied upon and that the development of course materials for distance education or online learning be based on an institution's fair dealing practices. The new exceptions in sections 30.02 and 30.03 are even more problematic: they offer exceptions for digitization of licensed materials that are so hampered and hedged that we will not describe them here.[13]

Publicly Available Materials (PAM) on the Internet

Section 30.04 creates a new exception for educational institutions and their staff to reproduce, publicly perform, or communicate to the public by telecommunication, for educational or training purposes, materials available through the Internet.[14] The exception does not apply where the material is protected by a technological protection measure,[15] or if the owner posts a clearly visible notice prohibiting such actions.[16]

During the years of discussion that led up to Bill C-11, the educational community was highly divided on whether to support or oppose this exception.[17] Many commentators, ourselves included, were concerned that it seems to presume that common and routine uses of the Internet in education are infringing, whereas *CCH* or theories of implied licence suggest otherwise.[18] In the end, in the light of the addition of education to fair dealing uses and the Supreme Court cases of 2012, the exception seems not particularly helpful and we hope it does not become harmful. It could perhaps

Who owns the copyright in students' work?
The student. Any reproduction of students' work beyond the bounds of fair dealing or other exceptions must be done with their permission.

Who owns the copyright in teachers' work?
It depends. Elementary- and secondary-school teachers often do not own work done in the course of their employment; to determine the policy, they will have to check their employment contracts. In community colleges, too, copyright often rests with the institution. In universities, faculty research has conventionally been owned by the faculty, but other work such as correspondence courses may be treated differently; academic staff, too, should check the collective agreement or university policy. Contracts with funders also sometimes contain constraints and conditions. See chapter 7 for more on copyright ownership and employment.

enable use of materials where the fair dealing tests seem to be somewhat equivocal, but again, it should only be seen as a backup for fair dealing, rather than a primary rationale for educational uses.

Licensing

Licensing with vendors is a major source of educational materials in schools, colleges, and universities. A growing proportion of resources are now digital and obtained through licences; in each case, there is a contract that sets out the terms and conditions of use. The terms of such licences are very much worth watching. In some cases they are very restrictive compared to the Copyright Act: they may limit the kind of archiving that researchers need, for example, or require that material be used on proprietary and often unsuitable platforms. This is something educational institutions are now in a position to push back on. Prices, too, are a major issue given near-monopolies in some areas of educational and academic publishing, and this pressure is producing major initiatives for Open Access educational resources.

In addition to individual vendor licences for specific materials, there are also general licences with various copyright collectives that provide blanket permissions for the various rights associated with works and other subject matter. For example, as discussed in chapter 6, Access Copyright has provided licences for photocopying, and Criterion and Audio Ciné have provided licences for film screenings in educational settings.

In recent years, the relationship between educational institutions and Access Copyright has become very controversial. In 2004 the Association of Universities and Colleges of Canada (AUCC) negotiated a licence with Access Copyright that was to last from 2004 through 2007. Most Canadian colleges and universities adopted this general licence, and it was not considered problematic at first. There were two forms of payment: Part A covered general unidentified copying using a rate based on the number of full-time equivalent (FTE) students at each institution (the rate was $3.58 per FTE by the last year of the contract). Part B assessed a charge of ten cents a page for identifiable copying, that is, the materials that went into coursepacks.

The 2004–07 licence was renewed for another year upon its termination for 2007–08, and again for 2008–09, and 2009–10. During the period of these extension years, there was growing dissatisfaction with the continuation of the licences. Many educators felt that after *CCH* the licence should have been renegotiated to reflect the increased importance of fair dealing. While the licence by its terms did not explicitly supersede or interfere with fair dealing (that is, copies of works amounting to fair dealing were technically excluded from the licence), the practical reality was a different story. On several campuses, the Access Copyright licences were presented as negating the need for fair dealing,[19] while the countervailing consensus grew that the permissions granted by the terms of the licence lay within the scope of fair dealing, and so should not have required payment.[20]

It is likely that the status quo would have continued for several years, but in March 2010, Access Copyright filed a tariff application with the Copyright Board[21] and the voluntary licences were not renewed. (For the difference between a licence and a tariff, see chapter 6.) The proposed tariff application would cover all post-secondary institutions for the period 2011 through 2013 (and as of spring 2013, it was still pending at the Board). Access Copyright was not merely trying to carry forward the terms of previous licensing agreements in the form of a Board-certified tariff. Rather, it was seeking a tariff with a higher cost per student ($45 per FTE at universities and $35 per FTE for the colleges, with page charges for coursepacks—Part B of the old licence—eliminated) and with a wider scope and broader application. In addition, the proposed tariff contained numerous audit, reporting, and monitoring provisions that would place new burdens on the institutions, their staff, and their instructors.[22]

While the highly contested Copyright Board proceeding was under-way, in January 2012, Access Copyright announced it had reached licensing agreements with the University of Toronto and the University of Western Ontario.[23] In April 2012, AUCC reached a similar agreement resulting in a model licence for use by other universities.[24] In May 2012, the Association of Community Colleges of Canada (ACCC) reached an agreement for a model licence for the community colleges. In the months that followed,

California has enacted new legislation that will create the California Open Source Digital Library to develop and house open source materials—that is, educational materials available for free to all. Despite strong opposition from the Association of American Publishers, the legislation passed the Senate and House with strong bipartisan support and has been signed by the Governor. The measures will

- establish the California Open Education Resources Council comprising faculty from the University of California, the California State University, and community colleges for development of a list of the fifty most widely taken lower-division courses
- require publishers, as a condition of a campus bookstore's purchase of specified books, to provide at least three free copies of the book to be placed on reserve at the campus library
- promote strategies for production, access, and use of open source materials
- develop a standardized, rigorous review and approval process for developing open source materials
- require that materials produced be placed under a Creative Commons attribution licence that allows others to use, distribute, and create derivative works based upon the material while still allowing authors or creators to receive credit for their efforts
- require that materials produced be modular to allow easy customization and encoded in such a format that they can be made available on the widest possible range of platforms and be accessible to persons with disabilities

In October 2012 the province of British Columbia announced plans for a similar initiative to create open textbooks with input from B.C. faculty, institutions, and publishers.

Source: Sam Trosow, "California Bill Would Create Open Access Textbook Digital Library," www.samtrosow.wordpress.com, 4 June 2012; "BCcampus to Co-ordinate Provincial Open Textbook Project," www.bccampus.ca, 16 October 2012.

several institutions signed on but many others chose to reject the model licences as unnecessary or unsuitable.[25]

In the aftermath of the enactment of Bill C-11 in June 2012 and the announcement of the Supreme Court decisions in July 2012, these licences are even less credible. In September 2012, ACCC advised its members not to sign its own model licence.[26] In the fall of 2012, several new sets of fair dealing guidelines for educational institutions were released, and it appears as if a consensus is finally being reached in the educational community that fair dealing is alive and well after all.

Even with strong fair dealing, educational institutions do have licensing and access needs. But any optimism that Access Copyright was going to shift gears and try to be more co-operative with educational institutions was dashed by the lawsuit it filed in Federal Court in April 2013 against York University, alleging that York's fair dealing guidelines authorize and encourage illegal copying. Access also announced additional filings with the Copyright Board in both the K–12 and post-secondary sectors.[27] So, after a brief moment of apparent stability, the copyright licensing situation in the educational sector is back in flux.

Fair Dealing Guidelines for York Faculty and Staff (11/13/12)

II. Fair Dealing Guidelines

1. Teaching Staff and Other Staff may copy, in paper or electronic form, Short Excerpts (defined below) from a copyright protected work, which includes literary works, musical scores, sound recordings, and audio-visual works (collectively, a "Work") within the university environment for the purposes of research, private study, criticism, review, news reporting, education, satire or parody in accordance with these Guidelines.

2. The copy must be a "Short Excerpt", which means that it is either:
   10% or less of a Work, or
   no more than:

(a) one chapter from a book;

(b) a single article from a periodical;

(c) an entire artistic work (including a painting, print, photograph, diagram, drawing, map, chart and plan) from a Work containing other artistic works;

(d) an entire newspaper article or page;

(e) an entire single poem or musical score from a Work containing other poems or musical scores; or

(f) an entire entry from an encyclopedia, annotated bibliography, dictionary or similar reference work,

whichever is greater.

3. The Short Excerpt in each case must contain no more of the Work than is required in order to achieve the fair dealing purpose.

4. A single copy of a short excerpt from a copyright-protected work may be provided or communicated to each student enroled in a class or course:

(a) as a class handout;

(b) as a posting to a learning or course management system (e.g. Moodle or Quickr) that is password protected or otherwise restricted to students of the university; or

(c) as part of a course pack.

5. Any fee charged by York for copying a Short Excerpt must not exceed the costs, including overhead costs, of the making of the copy.

6. Copies of Short Excerpts made for the purpose of news reporting, criticism or review should mention the source and, if given in the source, the name of the author(s) or creator(s) of the Work.

7. Where the Fair Dealing Exception allows the copying of only a portion of a Work, no member of the Teaching Staff or Other Staff may make copies of multiple Short Excerpts with the effect of exceeding the copying limits set out in Section 2 of the Guidelines.

For the full guidelines, see http://copyright.info.yorku.ca/ fair-dealing-requirements-for-york-faculty-and-staff

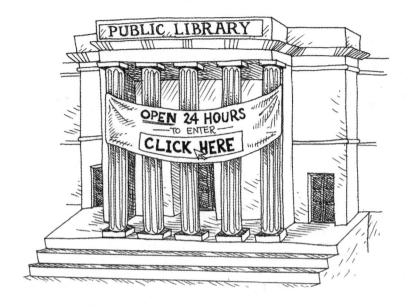

16. LIBRARIES, ARCHIVES, AND MUSEUMS

I n a democratic society, libraries, archives, and museums serve as storehouses of cultural heritage and history, as engines of public education (especially continuing or lifelong learning), and as promoters of public discourse on a range of issues. They collect, they select, they preserve, they catalogue, they reveal, they circulate. They often perform such functions with works whose copyright they do not own.

Reproduction, communication, performance, and exhibition rights are exclusively granted to copyright owners by the Copyright Act. By some visions of copyright principles, much of what libraries, archives, and museums do might seem like infringment. It is not new or surprising that a tension exists between some views of copyright law and the fundamental mandate of libraries, archives, and museums, known collectively in the

Copyright Act as "LAMs."[1] With the possibilities that digital technologies offer for copyright owners to control their products past the point of sale, on the one hand, and for LAMs to offer wider access to their collections, on the other, the tensions have increased in recent years. However, 2012 changes to copyright through legislation and judicial decisions offer support in the strongest terms for LAM activities. These public institutions now have the opportunity to focus not on fear but on what they do best: providing public access to a range of cultural and intellectual materials.

   This chapter dovetails closely with chapter 5 on Users' Rights and chapter 15 on Education. We will not go over in detail the importance of the *CCH v. Law Society of Upper Canada*, *Alberta v. Access Copyright*, and *SOCAN v. Bell* cases, because they are discussed at length elsewhere, but we start with fair dealing nonetheless as the foundation of LAM practice. After that, we proceed through a range of library activities and issues: collection management and maintenance, interlibrary loan, e-reserves, patron services, and pricing and contract terms for digital materials. Several library exceptions have been modified or added to the Copyright Act in 2012, and they are reviewed under these headings.

Fair Dealing

The basis for fair dealing in the library setting is the *CCH* case of 2004. In that case, explained in chapters 5 and 15, copying materials and faxing them to patrons was deemed to be fair dealing, and even more, a user's right. It represented a major victory for libraries, clarifying that

- fair dealing applied to libraries and their patrons regardless of the more specific and restrictive LAM exceptions;
- fair dealing was a much broader category than had previously been thought;
- librarians acting on behalf of patrons may "stand in their shoes" for purposes of helping them use their fair dealing rights; and
- the provision of freestanding photocopiers in a library did not constitute an infringing authorization under section 3 of the Act.

The *CCH* ruling provided a solid underpinning for a more proactive approach to library practice as well as to advocacy on copyright issues. Libraries have been overly cautious in applying *CCH* to their practices, but with the *Alberta v. Access Copyright* and *SOCAN v. Bell* decisions of 2012, their fears should be laid to rest. The *Alberta* case's language about the role of teachers in facilitating research amplifies the finding in *CCH* that a librarian stands in the shoes of the patron, and therefore is not infringing if the user is not infringing. Furthermore, the addition of education to the Act's fair dealing purposes will also provide protection for many if not all library patrons' uses.

Given these Canadian developments, making Canadian fair dealing closer to U.S. fair use, we contend that the U.S. case *Authors Guild, Inc. v. HathiTrust* concerning digitization of library collections is pertinent in the Canadian context. In September 2011, the Authors Guild and others (including the Writers Union of Canada) filed a copyright infringement lawsuit in New York Federal District Court against HathiTrust, the University of Michigan, and several other universities. HathiTrust is a coalition of U.S. research libraries that seeks to build a digital archive of library materials converted from print. The suit alleged that the libraries were infringing the plaintiff's copyrights by the unauthorized digitization of works in their collections. In response, the libraries argued that their activities constituted fair use, and the court emphatically agreed.

The defendants were using the digital files in three different ways: (1) full-text searches; (2) preservation; and (3) access for people with certified print disabilities. The full-text search capability allowed users to search for a particular term across all the works in the archive, but unless the work was in the public domain or otherwise authorized, the actual text was not returned to the searcher, just the pages on which the terms were found. Print-disabled individuals were given access to the collection through software that allowed the text to be conveyed in an audible or tactile manner.

The court rejected the plaintiff's argument that fair use was not available because of the special library exceptions in the U.S. Act. The court then turned to an evaluation of the four fair use criteria set out in section 107 of the U.S. Act (see chapter 5). On the first factor ("purpose and character of the use, including whether such use is of a commercial nature or is for non-profit educational purposes"), the court found for the defendants for several reasons. First, other than in the special case of print-disabled patrons, the full text was not generally being distributed. Second, the court found the search capability to be transformative because the copies served a different purpose than the original works did.[2] The court noted that the searching capabilities were already giving rise to new methods of academic inquiry such as text mining. Third, the court found that the provision of text services to print-disabled patrons was also transformative and therefore more likely to be fair.

On the second factor ("nature of the copyrighted work"), the court noted the general rules that some works are closer to the core of intended copyright protection than others, and that copying factual works is more likely a fair use than copying creative works. While some fiction was involved, the court said that because the use was transformative (searching and access for the disabled), the second factor was not dispositive.

On the third factor, the amount of the work copied, the court applied the general rules that the extent of permissible copying varies with the purpose and character of the use. The court said that in this case, it was necessary to copy the entire work in order to fulfill the library's purposes.

Finally, on the fourth factor, whether the use usurps the market for the original, the plaintiffs were arguing market harm on the grounds of lost sales. The court rejected this argument because purchasing another copy of the work would allow neither full-text searching nor access for the print disabled. The plaintiffs also argued harm to potential licensing opportunities, but this was rejected as conjecture.

The court had no difficulty in concluding that on balance the factors weighed strongly in favour of fair use:

> Although I recognize that the facts here may on some levels be without precedent, I am convinced that they fall safely within the protection of fair use . . . I cannot imagine a definition of fair use that would not

encompass the transformative uses made . . . and would require that I terminate this invaluable contribution to the progress of science and cultivation of the arts. . . .

This case is under appeal in early 2013, but the way the court goes through the fairness tests—essentially the same as those in Canadian law—is very important for Canadian librarians to consider as they pursue opportunities to digitize their collections. Note especially how it includes under fair dealing the reproduction of entire texts, questioning the self-imposed 10 per cent guideline that libraries often follow. It invites Canadian libraries to embrace fair dealing as a legal rationale for extensive digital sharing of their collections. But it offers something to creators, too: its distinction between creative and factual works may also be a line of argument to consider as the bounds of fair dealing continue to be worked out in Canada.

Management and Maintenance of Collections

The library exceptions in the area of collection management were amended in 2012. Section 30.1(1) of the Canadian Copyright Act provides that it is not an infringement of copyright for a LAM (or a person acting under its authority) to make a copy of a work (or other subject matter, such as a sound recording) for the maintenance or management of its permanent collection for six stated purposes. These six purposes apply regardless of whether the materials are published or unpublished.

First, under paragraph (a), in the case of a rare or unpublished original, a copy can be made if the work is "deteriorating, damaged, or lost," or at risk of becoming so. Note here that library staff no longer have to wait until the object is destroyed; they can now be proactive, as their profession demands.

Second, under paragraph (b), a copy can be made "for the purposes of on-site consultation" if the original can't be used because of its condition or because of the atmospheric conditions in which it must be kept.

Third, under paragraph (c), a copy can be made "in an alternative format" if the format of the original is considered to be obsolete or to be becoming obsolete, or if the technology required to use the original is unavailable or is becoming unavailable. Bill C-11 expanded the scope of this purpose to include situations where the format is *becoming* obsolete or where the requisite technology is *becoming* unavailable. Previously, the original had to already be in an obsolete format or the requisite technology unavailable. The amendment gives LAM staff the discretion to make decisions and is in keeping with the principle of technological neutrality.

The fourth part of the management/maintenance exceptions, paragraph (d), allows copies for "internal record-keeping and cataloguing." Since the Copyright Act itself does not define the term "cataloguing," it seems that the scope of the definition will depend on recognized library usage of the term. The fifth exception, paragraph (e), is copying "for insurance purposes or police investigations"; the sixth, paragraph (f), is copying for "restoration" purposes, where again, the ambiguity of the definition appears to leave the details to be filled in by LAM practice.

In addition to amending provisions specific to LAMs, Bill C-11 also added an exception for backup copies in either digital or analog form. While it is not limited to use in the LAM context, the new backup copy exception is of particular interest to LAMs and their staff. Section 29.24(1) provides that it is not an infringement for the owner or licensee of a source copy to make a backup copy under certain conditions. First, the reproduction must be solely for backup purposes in case the source copy is lost, damaged, or becomes otherwise unusable. Second, the source copy itself cannot be infringing. Third, in making the backup copy, a technological protection measure cannot be circumvented (see the discussion of the new anti-circumvention rules in chapter 8). Finally, the person making the backup copy cannot give it away. Notice that the last prohibition only applies to giving, not to lending. There is also a requirement that all reproductions made should be destroyed if the person making the copy ceases to own the copy or have a licence to use it.

Section 29.24(2) also provides that if the source copy is lost, damaged, or otherwise rendered unusable, one of the reproductions then becomes the source copy. On its face, this provision applies to LAMs and could prove useful especially for libraries wishing to make backups, particularly of material in sensitive media such as CDs or DVDs loaned to patrons. Unfortunately, since electronic media are often digitally locked, the potential use of this new section will be compromised by the prohibition on circumventing the TPMs (even for what in this case is otherwise a lawful purpose).

Overall, these exceptions provide LAMs with some safe harbours covering situations that typically arise in practice. The usefulness of the first three of these exceptions is limited to circumstances in which an appropriate copy is not "commercially available."[3] For this reason, it is important to think about fair dealing as a further justification. Commercial availability does not negate fair dealing the way it negates the special exceptions; under fair dealing some preservation activities may still be permissible, even if commercial alternatives are available.

Interlibrary Loan (ILL)

One tool that libraries have long used to make dollars stretch further and enhance their patrons' ability to perform research is interlibrary loan (ILL). A patron of one library can arrange through that library to borrow material from another library—perhaps in another city or another country.[4] Digital technologies can make ILL much easier: instead of sending the physical book, a library can fax or email a digitized version of the material. Questions have arisen about whether this practice constitutes copyright infringement. While librarians and patrons see ILL as a way of equalizing access to information resources across different communities, vendors and rights holders might see it as a threat to their income streams.

Bill C-11 sought to address these tensions through revisions to the existing exceptions. Section 30.2(2) now provides that it is not an infringement of copyright for a LAM (or a person acting under its authority) to photocopy a work from a periodical for any person requesting to use the copy for research or private study. In the case of a newspaper or periodical that is not scholarly, scientific, or technical, the exception explicitly excludes fiction, poetry, dramatic works, or musical works. Any copy from a newspaper must be more than one year old.

Note several limitations here. First, the provision is only applicable to works (unlike the previous exception that can apply to other subject matter, such as sound recordings)—and only some types of works at that. Second, it only applies in the case of research or private study. Now, while we know from our discussion of fair dealing that these categories are supposed to be broadly construed, it is not clear why this exception mentions research and private study but not the other enumerated fair dealing categories (education, parody, satire, criticism, review, and news reporting). Is it really the function of a library to interrogate patrons on their underlying purpose?

Before the 2012 amendment, section 30.2(5) provided that "the copy given to the patron must not be in digital form." Bill C-11 has added new language that has somewhat relaxed this limitation. But it is still subject to several restrictions and counter-limitations. Section 30.2(5.02) now provides that the LAM may provide the patron with a copy in digital form, but it has to take measures to prevent the requesting patron from

(*a*) making any reproduction of the digital copy, including any paper copies, other than printing one copy of it;
(*b*) communicating the digital copy to any other person; and
(*c*) using the digital copy for more than five business days from the day on which the person first uses it.

In other words, the library must now implement and attempt to enforce disabling mechanisms that prevent the patron from properly utilizing the requested materials. It is often necessary to print out more than one copy of an article (think paper jam in the photocopy machine, or simply misplacing the copy) and it is typical to want to share your work in progress with a colleague, collaborator, teacher, or student. Finally, it is often the case that you need to refer to the materials beyond the five-day period in which the work must somehow self-destruct. These requirements should not be imposed on libraries, their staff, or their patrons as they violate the principle of technological neutrality and force libraries to employ less-effective technologies than are otherwise available. For this reason, we cannot recommend relying on this section. Instead, libraries can frame their ILL policies, as all of their patron service policies, within their overall approach to fair dealing as articulated in *CCH* and *Alberta*. Doing so, they can focus on meeting their patrons' needs rather than on limiting their options.

E-reserves

Another challenge facing academic libraries is how to best organize course reserve materials. Traditionally, designated course materials have been maintained by libraries in paper form and held in closed reserve, requiring the student to present ID to obtain either the book or a photocopied file for a limited period of time. In practice, this system has been shown to be unworkable. Not only are unnecessary photocopies required, but the delay in getting access to reserve materials is a source of frustration for both staff and patrons, especially near exam time. One logical solution is to utilize digital technology to accomplish the same purpose, a practice referred to as "e-reserves." Despite the advantages of an e-reserve system over the old reserve desk model, many institutions remain nervous about implementing such a service because of copyright liability concerns.

The American Library Association (ALA) posted a useful guide to implementing e-reserve systems; they concluded that "while there is no guarantee that a practice or combination of practices is fair use, such certainty is not required to safely implement e-reserves."[5] ALA's assessment about the state of e-reserves and risk taking was vindicated by the recent ruling in a suit brought by publishers against Georgia State University in *Cambridge University Press v. Becker*. Although this is a U.S. case, we think that, like the *HathiTrust* case, it has value in the Canadian context in a number of ways.

This is how the case came about. In 2008, Cambridge University Press, Oxford University Press, and Sage Publications filed a lawsuit against officers of Georgia State University for copyright infringement. The publishers alleged massive copyright infringements resulting from the university's e-reserve system. Students would access the web-based e-reserve system by inputting a pass code obtained from the professor for the course. Pointing to its campus copyright policy, the university claimed the e-reserve system was fair use.

To determine when a particular use was fair, the court looked at the four statutory factors under section 107 of the U.S. Copyright Act. In 2011, the court found in favour of the university on the first factor (the purpose and character of the use) because the excerpts were all used for the purpose of teaching and because the university was a non-profit educational institution. While the publishers had relied on a line of cases (*Basic Books, Inc. v. Kinko's Graphics*, *Princeton University Press v. Michigan Document Services*, and *American Geophysical Union v. Texaco*) that found the first factor weighed against fair dealing, the court differentiated those cases because they concerned commercial copying. The court also found in favour of the defendant on the second factor (nature of the work) because in each case the excerpts were informational and educational in nature, and none were fictional. On the third factor (amount of the use), the court considered both the quantity and the value of the amount copied in relation to the overall book and whether the portion used was reasonable in relation to the work from which it was taken and the purpose for which it was used. Where a book was not divided into chapters or contains fewer than ten chapters, copying no more than 10 per cent of the pages in the book was considered permissible under the third factor. But where the book contained ten or more chapters, then copying an entire chapter was considered permissible.

On the fourth factor (effect upon the potential market), the court found that use of a decidedly small excerpt will not in itself cause harm to the potential market for the book because such an excerpt does not substitute for the book. However, the court said that if permissions were readily available (either directly from the publisher or from the Copyright Clearance Center) at a reasonable price and in a convenient format, then the fourth factor would weigh heavily against fair use. (With respect to the university's reserve system, permission would have to be available for digital excerpts.) But where such permissions were not readily available, the fourth factor would weigh in favour of fair use.

In the end, the court concluded that only five of the seventy-five excerpts submitted for decision constituted infringement. And after consideration, the court refused to issue the injunction requested by the publishers and awarded costs to the university. The court noted that the university tried to comply with the copyright laws and that there was no reason to impose the burdensome record-keeping requirements sought by the publishers.

The Georgia State ruling has several interesting implications for Canadian library practice. As we have already observed, the operation of fair dealing analysis has become very close to fair use analysis, with the inclusion of education as a fair dealing category and *CCH* and *Alberta*'s insistence that fair dealing categories be broadly construed. Canadian cases will turn on the factual analysis of the *CCH* fair dealing factors, which are very similar to the U.S. fair use tests. It is interesting to note that as in *HathiTrust*, the court in the Georgia State case distinguished between fictional and informational works under the "nature of the work" test. But on balance, the decision suggests that Canadian academic libraries can be aggressive in their development of robust e-reserve policies. It provides a good example of how a court might react to a technical infringement that is small in comparison to the entire scope of a school's copying. Even though the court did find a few infringements, it was very sympathetic to the university because it had put a copyright policy in place and attempted to use it. Furthermore, under Canadian case law, it is doubtful that even the five excerpts of the seventy-five would have been found infringing. Not only has the *CCH* court said that the availability of a licence is not relevant for fair dealing (*CCH*, para. 70), but also, at least at present, it would be more difficult for a publisher in Canada to show the availability of a convenient and reasonable transactional licence for e-reserves.

Patron Services

Libraries have historically been resistant to any sort of monitoring of the use patrons make of their materials: freedom to read is a fundamental principle. The collections of public libraries and museums have largely been accessible without payment or on an at-cost basis. However, LAMs are recently beginning to entertain the possibility that they ought to monitor or monetize how their patrons use their materials, equipment, or services. In this section, we look first at the monitoring of self-service equipment and platforms, next at copies of unpublished material provided to patrons, and lastly at fees for the use of materials from a collection. We believe that there are strong arguments—both ethical and legal—for libraries and museums to be as hands-off and generous as possible with how their collections are used.

In *CCH*, the publishers took the position that by making free-standing photocopiers available to its patrons, the defendant library was "authorizing" acts of infringement of its copyrighted materials and thus itself infringing the authorization right at the end of section 3(1) (see chapter 4). The Great Library of the Law Society of Upper Canada did not have a licence with Access Copyright, but it posted this notice above its self-service copy machines:

> The copyright law of Canada governs the making of photocopies or other reproductions of copyright material. Certain copying may be an infringement of the copyright law. This library is not responsible for infringing copies made by the users of these machines.[6]

The court ruled that the library was not liable for infringement because "a person does not authorize copyright infringement by authorizing the mere use of equipment (such as photocopiers) that could be used to infringe copyright" (para. 43). This is a statement that ought to resonate in libraries. And on the principle of technological neutrality articulated by the Supreme Court in *ESA v. SOCAN* (see box on page 136, in chapter 10), it should apply as well to more current equipment and platforms, such as scanners and course management systems on which copies can be made or posted. The *CCH* court was very clear that a library needs to articulate a policy, but it does not have to take on the job of copyright cop.

Special collections raise particular patron services issues as well. An amended LAM exception specific to archives, section 30.21(3), allows the making of a copy of an unpublished work for research or private study (but apparently not for education, parody, satire, criticism, review, or news reporting). The exception applies only if the copyright owner who deposited the work did not prohibit its copying, or if copying has not been prohibited by another owner of the copyright of the work. Also, under section 30.21(3.1)(b), the archive also must inform the requesting patron that the copy can be used only for research or private study and that any other purpose may require additional authorization. This exception essentially affirms common practice.

The last issue we will raise under the patron services heading is revenue generation or cost recovery. Libraries and museums are, of course, under budgetary pressure, and sometimes they may see their unique special collections or archives as possible gold mines. But there is a cost to the public interest in charging fees for access, for the right to take photographs, or for non-commercial use of collections. And if there is a cost to the public interest, there is also a cost to the reputation of libraries and museums. We strongly advise against such fees; while it is certainly appropriate to charge for large-scale commercial use, those within the community of learners or cultural workers should not be made to pay for access to our public domain. In her book *Permissions, A Survival Guide*, Susan Bielstein puts a humorous twist on the challenges of obtaining rights clearance for scholarly works including images: "Chasing images is time-consuming and expensive. So if you can live without images, do it. Put that money toward orthodontia or a down payment on a house."[7] Canada's libraries, archives, and museums would be wise to heed a warning provided by consultant Diane Zorich:

> Museums need to consider the ethical and moral dilemmas they create if they impose such restrictions as a matter of policy, for they would not wish to face similar restrictions imposed on them by others. . . .

Whether such restrictions are worth the price is an ethical question that museums must ask themselves when they consider access issues in their IP policies.[8]

Pricing and Contract Terms for Digital Materials

The broad diffusion of the Internet, including the widespread use of resources such as Google and Wikipedia, may threaten to undermine public awareness of the importance of libraries. Many people now find information at home, at their fingertips, and the information industry has, not surprisingly, responded to the revenue implications of these emerging markets. But not all people enjoy the benefits of networks and digitization to the same extent. The affluent can afford the latest computers, private premium high-speed Internet services, cloud services, training, entertainment packages, and subscriptions to digital databases, journals, and other media, while the less affluent remain dependent on the public library and slower connections for access to these services.

To meet the increased demand for online resources, libraries increasingly serve as portals for licensed digital materials rather than as the traditional repositories of purchased material. This shift makes them vulnerable not only to price increases by the vendors of these services, but also to licensing terms that circumscribe users' rights. Librarians now have to study—and recognize the importance of—the specific terms of licensing agreements, and do this quickly and well. For example, if a licence that allows subscribers to view but not copy the material can be had for half the price of a licence that allows copying, during a time when libraries are sorely pressed for funds, a library could be tempted to agree to such an arrangement. Sometimes users' rights are given up for pragmatic reasons, but the long-term cost certainly needs to be considered.

Another issue involves the question of preservation and continued access: many electronic resources are now marketed on what amounts to a rental basis. If a library terminates its subscription, patrons lose access to all of the materials previously provided unless the subscription terms permit perma-

Public Library Boycott of Random House E-books

The high prices of digital library resources are intimately connected with copyright issues. For example, the prices are higher than the prices of print resources based on the assumption that more users access the materials—but with education, research, and private study so bolstered by the courts, many of those multiple uses are fair dealing and ought not to require extra payment. Digital resources are often hedged in with digital rights management so as to make fair dealing impossible. Publishers tend to present terms for digital resources in a take-it-or-leave-it way—but perhaps libraries have to seriously consider leaving it if they ever want to get better resources and terms. They could take the lead of the South Shore Public Libraries, which undertook a boycott of e-books published by Random House due to the publisher's unaffordable pricing policies.

The SSPL, which serves several Nova Scotia communities including Lunenburg and Mahone Bay, posted an online petition for patrons to sign:

> SOUTH SHORE PUBLIC LIBRARIES provides readers access to a wide variety of formats, including downloadable ebooks.
>
> Recently, RANDOM HOUSE has drastically increased the price of ebooks for sale to libraries, sharply limiting the number of ebooks SSPL can purchase for our borrowers' enjoyment.
>
> Therefore, SSPL will stop buying ebooks published by Random House until they lower their prices for ebooks.
>
> We, the undersigned, ask Random House to lower their pricing for ebooks for sale to libraries to a level consistent with those they sell to individuals.

Imagine if university libraries could muster thousands of students to protest high subscription prices for academic journals, some of which notoriously cost thousands of dollars a year, even when their contents are generated within universities and given without payment to the journals. Or what if libraries rejected e-book licences that only allow one or two users at a time? With terms like that, they'd definitely be better off buying the physical book and scanning as necessary under fair dealing. Libraries have bargaining power, and ought to band together to use it.

nent archiving. A digital encyclopedia or magazine is not sitting bound on the shelf, and it can evaporate in an instant.

For university libraries, price inflation in subscriptions to scholarly journals, the so-called serials crisis, has become a major problem. The high prices of these journals are especially galling given that university faculty provide, select, and edit their content for free—and in some cases even pay page fees for publication. Judith Panitch and Sarah Michalak observe that commercial academic publishers "find themselves in the enviable position of selling research which they neither produced nor paid for to a high-demand market. They maintain an additional advantage because each journal title is a unique commodity, characterized by its specific focus and also by its prestige."[9]

To deal with this situation, libraries are using consortial purchasing—where a number of libraries band together to get a better price. Some librarians and their associations are also strong advocates for Open Access publications—in which libraries, educational institutions, or other non-profit entities host journals that are made available at no charge to the end-user.[10]

In recent years, new challenges for collections budgets have arisen because of the increased demands for e-books.[11] Here, too, librarians will have to monitor contracts for their acknowledgement of users' rights along with other practical, price, and presentation factors.

PART IV

CONTEXTS

17. COPYRIGHT'S
COUNTERPARTS

Copyright is often a disappointment or an irritant. People often wish that it would allow more money to flow in the right direction—or more creative juices, more respect, or more knowledge. There is no doubt that it is very technical and financially unavailable to many of us. Even if we have good legal advice, copyright can make us itchy or restive: it often doesn't seem to match our sense of what is *right*. For example, a video artist who makes less than $15,000 a year might ask, Why should I pay hundreds of dollars for a piece of Apple software when I already bought a two-thousand-dollar Mac, I'm earning so little, and I can get it for free? Or a poet might ask, How can it be fair dealing for a teacher to copy my poem for free for her whole class? Or an artist might ask, Why is my work protected as fair dealing if it criticizes another work or a cultural phenomenon, but if it's a pastiche in tribute to other work, it could be copyright infringement? Or a scholar might ask, Why are dead authors' unpublished works protected by copyright for much longer than their published works?

In this chapter, we wish to acknowledge the important fact that copyright does not sit easily with the ways that many of us think about creativity and culture much of the time. This should not be a huge surprise. All law is a formalization of certain rules that often parallel—but sometimes diverge from—moral and ethical codes that underpin human conduct. If you refrain from killing your neighbour for letting her dog leave souvenirs on your lawn, it's probably not because you're afraid of the police, at least not directly. It is more likely because you were raised to think of murder, however apparently well deserved, as an act of evil. If you drive a bit above the speed limit on a straight, clear stretch of road, you are also following non-legal common sense. Neither of these examples suggests that we should not have laws—against murder and speeding, in these instances. But they do suggest that the legitimacy of the law depends on people generally feeling that it lines up with what they think is right and doesn't impose itself too visibly on their ordinary transactions and decisions.[1]

When drafters were coming up with Bill C-11, they no doubt had such issues in mind. Time-shifting of recorded TV shows, for example, is just something people do. Making it illegal would probably produce more harm to the law than benefit for rights holders. So they acknowledged an exception for it (see Table 8). Similarly, allowing an eleven-year-old to make a stop-motion animated movie with her American Girl dolls using music available online fits most people's ideas of fairness, and doesn't seem to harm rights holders very much. So they made an exception for non-commercial remixing (chapter 10). And it may just have seemed reasonable to give performers moral rights (chapter 4) and to treat photographers the same as other creators (chapter 11), other changes made in Bill C-11.

Justice Abella of the Supreme Court also appealed to common sense when she put herself in the shoes of the teacher in *Alberta v. Access Copyright* (2012) and invoked a "reasonableness standard" (para. 37) for a defence of the making of multiple copies for classroom use (chapter 15). And in the unanimous decision in *SOCAN v. Bell* (2012), the tone of Justice Abella's ruling is somewhat long-suffering in response to the cleverness of SOCAN's arguments that consumers should pay for access to clips of tunes available to promote online sales (chapter 9); ultimately she boils the whole issue down to the Copyright Board's claim that "short, low-quality streamed previews are reasonably necessary to help consumers research what to purchase. I agree" (para. 46). The law often works best when it approaches common sense.

One problem, though, is that in a complex world there is more than one version of common sense. Furthermore, copyright at its very core has its drawbacks. It doesn't always perform well when it comes to policy objectives such as freedom of expression, broad participation, and preservation. As a monopoly mechanism, however tempered, copyright tends to limit access and constrain use. As a market mechanism, copyright privileges market value over other ways of thinking about what is important. It rarely generates the most revenue for the best work or for the most labour. As a cultural policy mechanism, copyright provides an incentive for some kinds of creativity and dissemination, but not for others. From the point of view of fostering a vibrant culture, there are better and worse versions of copyright, but they will always have limitations. We need, then, to be aware of the alternatives that exist all around us. Copyright is not the only tool we have.

In recent years, some legal scholars have become quite alert to alternatives to intellectual property, producing studies of, for example, how French chefs, U.S. stand-up comics, entertainment magicians, and even roller derby girls practise and articulate norms of appropriate and inappropriate borrowing and acknowledgement of each other's work—norms that sometimes do and sometimes don't line up with copyright.[2] Boatema Boateng's excellent book *The Copyright Thing Doesn't Work Here* (2011) amply shows the misfit between Ghanaian textile practices and intellectual property law. In this chapter, we discuss what we consider two of the weightiest and most established alternative systems to copyright: Indigenous cultural property protocols, and citation economies. We also describe a well-developed alternative way of hacking copyright law, or adapting it to alternate values and working practices: open source or "copyleft." Finally, we discuss public funding, which in allowing measures of value other than the market, is an alternative mode of promoting creativity and knowledge production, and one with a strong Canadian tradition.

Knowing about these other ways of thinking and doing, and granting them conceptual and practical weight, is an important context for what can become very narrow copyright catfights. Each of these alternative mechanisms is in some ways or in some contexts capable of promoting policy objectives usually associated with copyright—that is to say, incentivization and regulation of creative production. These economies of knowledge are not pie in the sky: they are thriving now, and indeed several of them are much older than copyright. They challenge some of copyright's fundamental principles, but they are not incompatible with some forms of copyright law. In fact, copyright will really only work optimally in concert with such counterparts, so it always needs to leave space for them to thrive.

Indigenous Cultural Property

Although the recent increase in copyright controversy is often ascribed to the growth of the Internet and digital technologies, older technologies such as storytelling and traditional medicine also pose a challenge to intellectual property law. Over the past several decades, Indigenous people around the world have begun to coordinate their efforts to resist appropriation of their cultural knowledge and to assert other modes of governing its circulation. They have been pushing their case at local, national, and international levels.[3]

Some of the motivation is economic: Indigenous cultural knowledge has commodity value and, in Canada, may be one basis for sorely needed economic development in First Nations communities. Inuit art co-operatives may be one model here—even though most of the money in the global Inuit art market does not flow north. But Indigenous traditional knowledge (TK) has other sorts of value as well, value that motivates many of those advocating its special treatment. It has been developed collectively by generations of people, building up community, sacred, ecological, and political value along the way. Indigenous groups have difficulty effectively protecting these kinds of value through the copyright and patent systems, because these systems place TK in the public domain, where it may be harvested by outsiders for financial gain or artistic glory. Because it is old, unfixed, and has no single author, TK is an awkward fit within the conventional scope of copyright. In Canada as in other countries, Indigenous groups and individuals are working out ways of articulating their own laws and practices governing the circulation of cultural objects and images to those outside their communities.

In the light of copyright law, some Indigenous statements on copyright are reminiscent of the spirit of moral rights: they warn against reproducing or altering works in such a way as to produce damage to honour or reputation. The honour or reputation of concern here is not the author's, but rather that of the clan, culture, or nation. And indeed, many Indigenous people emphasize that the "author" of a specific expression is a tradition-bearer, not an originator. It is the tradition-bearer's responsibility to transmit what has been entrusted to him or her to the right people in the right way and at the right time. Depending on the particular tradition, the (re)creator may not be free to reinterpret the material in a new way, to disseminate it to strangers, or to sell it.[4] Thus, while alienability is foundational to Western

> Within our customary law only a clan member can wear or use anything carrying the clan crest. The clan, not an individual member, can give others the right to use the clan crest. The right to use the crest is generously given out, but only when requested in a good way, only for a good purpose and only for the narrow purpose requested. For example, we might allow an artist to use the crest in a carving, but would not sell the crest symbol to adorn a Coke bottle.
>
> — Christle Wiebe, Carcross/Tagish First Nation, "Customary Law and Cultural and Intellectual Properties."

ideas of property and intellectual property—you know you own something if you're allowed to sell it—Indigenous ownership, as many explain it, is based on ideas of custodianship, community, and responsibility.[5]

Tensions often exist between Indigenous and non-Indigenous approaches. For example, non-Indigenous poet and scholar Robert Bringhurst has long been devoted to celebrating and publicizing Haida culture. Between 1999 and 2001 he published three books of retranslations of hundred-year-old transcriptions of Haida material. Bringhurst was eager to induct Haida storykeepers Ghandl and Skaii as poets into the "polylingual canon of North American literary history," and one of the books brought Bringhurst a nomination for the Governor General's Literary Award.[6] But some considered his work appropriation. As Jusquan, a young Haida writer, put it in the magazine *Redwire*:

> These Haida stories have been mysteriously defined as "Canadian" by those who write book reviews and teach Canadian literature courses, as well as those Canadians and others who are buying those books. . . . The richness and complexities of Haida history and spiritual beliefs are now nicely understandable and transparent for Canadians and other non-Haidas, and oh isn't it all so quaint and glorious. Now there is an assumed ownership of our stories, not only by Bringhurst but by those who read his books, and there was no payment. And by payment I mean in recognition of the significance of those stories, or in the recognition of the living Haida today and what those stories signify and mean to the on-going struggles of our culture.[7]

To some readers, this anger may seem extreme. But consider the abundant totem pole key rings and raven T-shirts at the Vancouver airport, or the (astounding) choice of a Gumby version of the Inuit inukshuk as a logo for the 2010 Vancouver Olympics. The irony with the Vancouver Olympics logo is that the Olympic Committee went so far as to arrange for special legislation to fortify its rights in this appropriated image. Canada does help itself to Indigenous culture on a regular basis.

Although Indigenous protocols are specific to their various nations, they also have wider importance. These protocols are, after all, part of Canada's colonial past and present. In nineteenth-century Canada, settlers were given land just for showing up and cutting the trees down on tracts appropriated from Indigenous people; in twenty-first century Canada, settlers can stake claims on First Nations cultural materials that Western law claims for all Canadians. It's an uncomfortable parallel, and it ought to be addressed. Some creators are concerned about limitations on freedom of expression that might result if increased authority is accorded to First Nations cultural protocols. But the copyright system works by imposing exclusion mechanisms on instances of culture; while these mechanisms constrain the actions of users of this material, we accept them if on balance their effect is judged to be productive. This should be the principle by which we explore the place of First Nations cultural property within or alongside Canadian IP law.

There are hundreds of Indigenous nations in the Americas, and every one of those nations has customary laws that regulate how their knowledge is used and accessed. These legal regimes are at least between five thousand and twenty thousand years old. The great irony about the status quo today is that Western intellectual property rights are only two or three hundred years old at the most!

And there's the other irony, that is being told that Indigenous legal regimes are not legitimate and that we have to adhere to this so-called universal system, which came from a very small part of the world called Europe. The intellectual property rights system was imposed on Indigenous knowledge systems without the consent of Indigenous nations, and the conflict is one of legal regimes. It's a legal power play; it's an unjustly and immorally applied conflict between laws and sets of laws.

— Greg Young-Ing, Chair, Indigenous Peoples' Caucus, Creators'
Rights Alliance, Vancouver, interview with Laura Murray.

Beyond the political imperative, Indigenous ideas about cultural property function as a powerful critique of certain elements of copyright thinking: they illuminate a range of different cultural practices and contexts. For example, copyright does acknowledge collective creation in provisions for joint authorship, corporate ownership, anonymous publication, and the resource of the public domain, but at its heart is the fiction that the individual creator can make something alone. Some critics would say that Indigenous protocols, by downplaying the creative initiative of individuals in favour of a collective, inherited foundation, go too far in the other direction. But at the very least, this emphasis is a potent corrective.

Citation Economies

There is a widespread tendency in today's media and water cooler chatter to conflate "copyright infringement" and "plagiarism." True, they both constitute inappropriate borrowing of the work of others. But they are significantly different infractions, and they take place in two distinct economies of knowledge: copyright and citation. The fundamental expectation in the copyright system—notwithstanding users' rights—is permission, but the fundamental expectation in a citation system is attribution. Hence copyright infringement is use without permission (again, with all the caveats mentioned in chapters 3 and 5), and it's a matter of law, whereas plagiarism is use without attribution, and it's a matter of community or professional practice.[8]

Beyond the practical headaches caused by the confusion between copyright infringement and plagiarism, we see it as a disturbing indicator of society's lack of awareness of the workings and vitality of "citation economies." Citation tends to be associated with academic scholarship—and we'll get to that realm shortly. But listen closely: people can't even talk about the weather without quoting each other. To mark the authority or tone or social value of these quotations, ordinary conversation often requires citations as well. "Who told you that?" your friend asks—she wants a footnote. Or "Where'd you get that lick?"—musicians want to trace their lineage as they reinvent it. In their everyday lives people seem to understand the idea that, as language theorist Mikhail Bakhtin put it so well, "the word in language is half someone else's. It becomes 'one's own' only when the speaker populates it with his own intention, his own accent, when he appropriates the word."[9]

In a citation economy such as conversation, improvisational music, or the blogosphere, you don't need to get permission—you just need to acknowledge your sources. What these systems may lack by way of direct financial incentives, they make up for in other ways. Citation economies work for a number of different reasons. While some people speak of citation economies as "gift" economies, this useful analogy should not be taken as a licence to view them through a Hallmark haze. Citation economies build on individuals' desire for recognition and communication. We quote others to call them to account, to bolster our authority, or to add zest to our communication, and we deeply hope that others will do the same with our contributions. Policing happens not through law but through social expectations—plagiarism is seen as an insult to the group's honour and legitimacy. The vast majority of human communication and knowledge production emerges out of such environments.

Academic research, along with its junior siblings in schoolroom citation practices, is the most formal and established of the widespread citation

Why do citation economies work?

- no friction of time, money, effort, rejection risk in seeking permission
- less vulnerable to censorship than copyright
- effective at tracing origins and relationship of ideas
- effective at moving information and expression fast and far
- effective at innovation
- appeal to altruism
- appeal to egotism
- build community
- allow social differentiation and status markers
- lead to opportunities for remuneration

economies. In the "web of science,"[10] each person's writing is understood as a contribution to a network of dialogue and expertise. Since currency in this citation system is reputation, barrier-free access and circulation are essential. Today this citation economy is under considerable pressure. Universities give away publicly funded knowledge to for-profit journals and buy it back, if they can, at high prices. In agreements with private funders, or in an effort to generate income, they constrain access to research data or resources that were formerly broadly available. Out of fear of liability, they have been paying licensing fees for uses of works that are customarily and often legally acceptable.

These constraints are not only happening at the post-secondary level. School boards have been fed the mantra that it is their responsibility to educate children about copyright.[11] Doing so is, in our view, irresponsible—at least in the junior grades. If children get the message too young that they must ask before they can use anything around them, they will become obedient consumers before they have the chance to become fearless learners or budding creators. Let's get our priorities straight. People create because they want to—and they do this long before, and often long after, they are motivated by money. After we are sure that students have a sense of engagement with their cultural and intellectual heritage, and a sense of participation in its development, then we can tell them about copyright. Copyright has a place in cultural development, and it's a secondary one. Happily, recent legislative and judicial developments give students and teachers the space they need to learn without clearing permission at every step.

We recognize that outside of academe, few can make a living off being quoted; as artist John B. Boyle said, "This is Canada: you can die of exposure."[12] We are by no means suggesting that the copyright system be replaced by citation modes—or copyleft or Indigenous models, for that matter. Our argument is simply that crowding these systems out will impoverish us all—culturally and indeed, ultimately, financially.

Open Source Knowledge

Unlike the long-established economies of knowledge arising in Indigenous and scholarly contexts, open source (OS) has emerged fairly recently out of computer programming to become a model and inspiration for many knowledge-sharing and knowledge-building projects. In a world where we have usually presumed that we have to pay people to create value, OS poses a rather exhilarating alternative. Perhaps we can let Wikipedia, a now-ubiquitous resource based on OS principles, define the term:

> In production and development, open source is a philosophy, or pragmatic methodology that promotes free redistribution and access to an end product's design and implementation details[*citation needed*] . . . Generically, open source refers to a [computer] program in which the source code is available to the general public for use and/or modification from its original design free of charge, i.e., open. Open source code is typically created as a collaborative effort in which programmers improve upon the code and share the changes within the community. Open source sprouted in the technological community as a response to proprietary software owned by corporations . . . A main principle and practice of open-source software development is peer production by bartering and collaboration, with the end-product, source-material, "blueprints", and documentation available at no cost to the public.[13]

Notice the parenthetical "citation needed" note in this passage: it indicates first how important citation is to this system, and also the way that open source projects reveal their collaborative process on their face.

Open source software operates under what is known as the GNU General Public License (GPL). Programmers working under the GPL retain the right to be recognized for their contributions, but do not hold exclusive economic rights and cannot block reuse of the work. In provisions that have come to be known as "share-share-alike," subsequent generations of innovators must adhere to the same type of licence.[14] Later programmers need no

Australia has been in the forefront of developing cultural protocols for the circulation of Indigenous cultural materials. Protocols are voluntary codes of behaviour that in a sense work by peer pressure: they are being adopted and developed by more and more Indigenous communities, libraries, and museums worldwide as part of a huge effort to foster more ethical use of Indigenous materials than has prevailed in the past. Now, drawing on the Creative Commons model, Mukurtu CMS has launched "a free and open-source community digital archive platform powered by drupal that delivers professional standards-based archiving specifically designed for Indigenous digital heritage." It streamlines the permissions process for both users and creators inside and outside the community, and helps Indigenous knowledge enter the wider world in appropriate and proactive ways. And it offers an exciting prospect of collaboration between open source and Indigenous modes of cultural production and custodianship, both of them quite distinct from copyright.

Resource: Australia Council for the Arts, "Protocols for Producing Indigenous Australian Visual Arts," 2nd edition, 2007; www.mukurtu.org.

permission to use, improve, or customize the program, but cannot commercialize it: they must pass it on. This is indeed a very different protocol from copyright. Despite a lack of the profit incentive that economists assert as a necessary incentive for creativity, and despite institutional and market pressures to limit the spread of free software, the system has clearly thrived.[15] It has inspired a wide range of projects and lines of thinking, going under names such as "copyleft," "free culture," "Open Access," "Access to Knowledge (A2K)," and so on. These projects use copyright against itself: they take the rights the Copyright Act grants to authors and deploy them differently.

In 2001 the GPL model was adapted for a wide range of other creative practices through the Creative Commons licence, launched by Lawrence

> Today, as the Internet and the digitally networked environment present us
> with a new set of regulatory choices, it is important to set our eyes on the right
> prize. That prize is not the Great Shopping Mall in Cyberspace. That prize is
> the Great Agora—the unmediated conversation of the many with the many.
>
> — Yochai Benkler, "From Consumers to Users," 565.

Lessig and others at Stanford University. Canadian versions of the licences appeared in 2003, and they have now spread to over seventy jurisdictions around the world. In essence Creative Commons licences allow a creator to choose which rights from the copyright bundle to reserve, and which to waive. Or, as the Creative Commons website puts it, "We use private rights to create public goods."[16] Rights holders have a choice of a range of different licences. All require attribution, but some permit non-commercial use only or require that the work not be changed. If creators stipulate that the derivative work be licensed on the same terms as its source—offering a "share-alike" licence—they are fostering a copyleft economy of circulation by perpetuating the terms of sharing. One of the important and sometimes unappreciated features of these licences is that creators who choose them do not have to forego financial reward: users who want to work with material in ways not permitted by the licence negotiate with the rights owner to do so, just as they would in mainstream copyright. Thus a song might be available for free non-commercial sampling, but Justin Bieber would pay for the rights if he puts it on his next album.

Over only a few years, Creative Commons licences have been taken up not only by individuals but also by prominent educational projects—the Public Library of Science (PLOS) and MIT's Open Courseware program are prominent examples. Open Access academic journals use similar sorts of licences to make research available for free, while protecting its authenticity and integrity. Specific licences have become a way of marking a certain vision of the Internet, in which access to information goods, their transformation into something new, and their redistribution to another creator-in-waiting is experienced as an ongoing process.

While, in its strong streak of libertarianism, much copyleft or free culture thinking contrasts with the Indigenous viewpoint (copylefters celebrate unfettered appropriation, while Indigenous people often speak against it[17]), the two approaches have points in common insofar as they are both based on an idea of responsibility for shared culture. They both strive, for example, to serve community goals, to use cultural engagement to build community, and to recognize in custom a kind of law. They both ask us to think twice about how we use other people's creations, rather than simply waiting for the law to tell us what to do.

Public Funding

Another counterpart to copyright can be found in public support for education, science, research, broadcasting, and the arts. Government funding has been an essential motor of cultural and intellectual production in Canada. But what is government funding? If we speak instead of taxpayer, public, or citizen funding, we might better evoke the responsibility that funded institutions and creators have to the public. Nowadays copyright and other intellectual property mechanisms are being held up as a great hope for funding education, research, and the arts. This is a kind of offloading. It asks cultural workers to bear alone the huge costs of training, development, and marketing of their work, for an exceedingly uncertain financial outcome even then. It produces unconscionable access barriers: only those with deep pockets or well-appointed basements can afford to indulge in creative careers. Along with increasing dependence on private-sector funding, it shifts the products and results of this sector to the detriment of Canadians. Although IP has a role in cultural and economic growth, it can't do the job alone.

The nation's well-being and future success require proper levels of funding for educational institutions and public libraries. These institutions are the bedrock for an informed citizenry. They provide future creators with

We asked Amanda Crocker, the Managing Editor of Between the Lines, how she sees the current situation for small publishers like hers. She didn't pull her punches:

> I have mixed feelings about copyright. I don't think big corporations should make as much money as they do when selling music, movies, or books. I don't think Disney should be able to have ownership of Mickey Mouse forever. I don't think ideas should be owned by corporations. But I'm also annoyed by people who have come to expect cultural products for free and forget that many people work to create these products and rely on payment for their work in order to continue producing such work. We can't work on editing, designing, printing, marketing, promoting, and distributing books and then just give them away for free chapter by chapter.
>
> Yes, we are funded in part by government grants (at least for now) but so are the automobile manufacturers and no one is arguing that part of their car should be free. Where is my share of the subsidized oil? If people no longer want to pay for cultural products then we all need to work together on a dramatically different economic model. Otherwise independent publishers like BTL won't survive. Some would argue there is no need for publishers in a new world where self-publishing is possible. Having worked with many authors who needed our help to focus and shape their ideas, to make sense of snarled sentences and logic problems, to organize jumbled research, I disagree.

We disagree, too. Writers need publishers, and readers need publishers. And it's important to remember that Canadian publishing has never been a free-market proposition. It is in the interest of all of us to maintain and adapt ways of supporting it in the present legal, cultural, and technological environment.

the intellectual tools needed to go about their work; they nurture the audience and indeed the market for creative work. Their universal accessibility is a public good of the highest order. But accessibility is not only about funding levels, it is about attitude. Schools and libraries could do much more to multiply the effectiveness of the funding they do get by sharing the fruits of their labours as widely as digital technologies now permit. Generous support for public broadcasting, performing arts organizations, and museums and galleries is also essential. In this regard we can clearly see the costs of cost-recovery strategies: when the CBC charges a documentary filmmaker $187 for one second of footage, or a gallery charges multiple layers of user and clearance fees for public domain images, it is preventing Canadians from benefiting from resources they have already paid for.[18] In their dealings with government, these institutions continually pronounce their indispensability—but perhaps they would see better results if they focused on a new or renewed commitment to serving the public and valuing the contributions of creators. As John Holden declares, "If a sustainable base for culture is to be secured then cultural professionals need to think of 'advocacy' not just in terms of generating 'evidence' for their funders, but as establishing broad support with the public."[19]

A good place to start on public relations is to make digital archives and collections available free for non-commercial use—and on a sliding scale for revenue-generating use. The National Film Board is moving in this direction. Free and sliding-scale admission prices fit this philosophy well, too. Of course, these sorts of policies and projects cost a great deal of money—but we believe they are the only way of reversing the slide in government funding. In a strange but compelling paradox, arts, research, and broadcasting leaders could demonstrate their indispensability more dramatically than ever before by giving away the shop—emulating Google, MIT, and other leaders in the Open Access movement. They could also improve their image with government and private funders by showing generosity to creators, and working to change the adversarial relationship that often seems to exist between management and content providers.

Direct funding to cultural workers and researchers is also an essential part of Canada's intellectual and cultural economy. At least in theory, granting councils don't pay out for what is popular, but for what juries of peers think is important, beautiful, or provocative. In this way they support currents of ideas and art that may not generate immediate income, but that will make people see new things or scratch their heads. The Public Lending Right, a program of the Canada Council, pays authors whose books are found in libraries to reflect dissemination beyond sales. Other possibilities could be pursued: for example, targeted tax policy to acknowledge the value and precariousness of artistic labour.[20] Such mechanisms have value both in themselves and in the economic spinoffs they produce as creators and audiences alike develop and diversify.

There are accessibility issues in this area as well: publicly funded work ought to be widely available. Academic researchers can publish in Open Access journals or resources and thus reach beyond the walls of their discipline or institution. Artists can make some of their work available for free.

To emphasize a responsibility to the public is not to say that all publicly funded art or research ought to be widely popular, widely understandable, or widely approved. Some art is potentially offensive or altogether perplexing, and much research is hedged in by biochemical or sociological jargon. So be it. The principle is that such work must be available without undue constraints to those members of the public who want to engage with it. In a world dominated by market ideology, "cheap" is a derogatory term and "free" may sound like garbage or trickery. But as many corporations and collectives are discovering, giving things away can be an enriching strategy.[21] There are costs and arrangements to be worked out: creators of materials in public collections may need to be paid extra for the digital use of their work, and in some cases such as certain Indigenous work, limitations on access may be necessary. Fees for commercial reuse will be necessary and in some cases might be increased. But with a mixed and flexible approach, the educational, research, and cultural sectors would be better positioned to fulfill their democratic mandates.

18. COPYRIGHT'S FUTURE

One might suppose that after the passing of Bill C-11 and the Supreme Court quintet of cases in 2012, neither Parliament nor the Supreme Court would wish to return to copyright very soon. Certainly the professional, cultural, and consumer groups that have devoted so much energy to the topic might welcome the chance to attend to other pressing issues. However, Canada didn't get much of a copyright holiday.

In March 2013 the government tabled Bill C-56 (the Combating Counterfeit Products Act), which would amend both the Trade-marks Act and the Copyright Act to provide new criminal offences and enact new border enforcement measures. The same month also saw an appeal launched in the *Warman v. Fournier* case we discuss in chapter 4. In April 2013, Access Copyright filed a lawsuit in Federal Court against York University alleging that York's fair dealing guidelines authorize illegal copying and asking that

York be ordered to operate within the Copyright Board's Interim Tariff. On the same day, Access also filed a request for the Copyright Board to force K–12 schools to comply with a tariff even though (or rather, because) the schools have determined that they do not need a licence from Access Copyright. And as a third salvo, Access filed a proposed post-secondary education tariff with the Copyright Board for 2014–17 mirroring the earlier application for 2011–13. At the time of publication, it is still pending at the Board. Meanwhile, in the United States, the Authors Guild appealed the trial court decision in *Authors Guild v. HathiTrust*, which we discuss in chapter 16.

These actions represent forceful pushback from owners' rights interests against recent gains on the users' rights side—in fact, such contestations confirm the importance of those gains. We have no crystal ball on how any of these cases will unfold; they all need to be watched closely, along with a broader suite of issues and areas of practice.

Collectives: Access Copyright and SOCAN have already jumped into the legal fray again even after the Supreme Court slapped them down. More generally, will copyright collectives continue to sue and otherwise attempt to impede legitimate uses of copyright material by user groups? Or will they find new products to offer that better suit the market? For example, educators, website designers, and consumers all seek convenient access to high-quality streaming video and digital versions of print materials. Instead of threatening to sue over the scope of fair dealing, perhaps collectives could follow models such as Spotify, iTunes, and the like, providing licences or downloads for low fees. Those models show that people will pay for access, even for materials available for free elsewhere. Or maybe we will see the emergence of new collectives, if the old ones can't keep up with the times. Collectives can be a very good way to manage rights, but they certainly face and pose challenges at present.

Universities and Colleges: Will post-secondary institutions continue to move towards adopting robust fair dealing policies, or will Access Copyright's April 2013 barrage of legal moves restore the previous state of inaction and risk aversion? Will the educational institutions back up their claims about being guardians of the world's knowledge by sharing that knowledge widely through Open Access digital repositories? Or will they resort to proprietary systems of online courseware in an attempt to raise revenues in tough times? Canada could become a leader in digitizing archives and otherwise sharing its intellectual and cultural wealth. Yet it remains unclear whether that will happen.

New Business Models: Will new models for distributing creative work really emerge? And to whose benefit? For example, income from copyright is more a mirage than a reality for most musicians. People say they can make money by touring, but this is only true for a vanishingly small higher echelon. What other income sources can be tapped or created? As more film and print material becomes available for unauthorized download, will people working in those media be able to afford their chosen career? Will government step in and offer more support for Canadian artists, writers, publishers, galleries, museums, and cultural innovators?

Digital Rights Management: Will the cultural and software industries continue to use digital rights management? Will the DRM protections enacted in 2012 block fair dealing, or will regulations be developed, as in the United States, that mitigate their harm to users' rights?

International Treaties and Trade Deals: Negotiations for the Trans-Pacific Partnership, a multination trade deal with IP components, have been conducted in secrecy such that citizen response and input are impossible. A leaked draft text does not manifest a balanced approach to copyright: users' rights are entirely absent. The International Intellectual Property Alliance (IIPA), an umbrella group for music, film, and software industries, is hoping that the TPP will undo provisions in Canada's copyright law such as low statutory damages and consumer exceptions for time-shifting and remix, and at the same time that it will lengthen copyright term and increase criminal penalties and enforcement. IIPA has access to the negotiations via a U.S.

regulatory process, whereas public interest voices like the International Federation of Library Associations and Institutions (IFLA) are limited to press releases from the sidelines. This situation is unfair and must be insistently critiqued. The introduction of Bill C-56 is a warning that we must also watch out that Canada does not roll over to U.S. pressure in implementing the highly flawed Anti-Counterfeiting Trade Agreement—or any future treaties with which the United States will try to impose imbalanced copyright on other countries.

Legal Reform: Bill C-11 requires a review of the Copyright Act in 2017. Issues that still require consideration remain: reform of Crown copyright, limitations on the waiver of moral rights, resale rights for visual artists, and new policy for orphan works are among them. Sanctions for misuse of copyright are another matter to look into, as are reforms in the area of copyright collectives.

Notwithstanding ongoing challenges and uncertainty, we close on an optimistic note. When we look back at the past ten years of copyright debate and developments, we do feel that Canadian copyright policy is getting closer to the tool it needs to be. Much progress has been made in turning Canada into a model for fair copyright practices and a jurisdiction that other countries can look to for guidance. The courts have shown serious concern for balance of the various interests at play. So has Parliament. But for the long-term health of copyright in Canada, what is most important is that so many people have become engaged with and informed about a fairly complex and often intimidating part of the law. Any new developments will face scrutiny from many directions, and this is as it should be.

Notes

For further detail about the sources listed in these notes, please refer to the Bibliography and Legal Cases Cited.

1. COPYRIGHT'S RATIONALES

1   Milton, "Areopagitica" (1644). The anonymous author is quoted in Rose, *Authors and Owners*, 55. Rose notes the possibility that the author of this pamphlet was in fact a bookseller—and therefore using the "authors' rights" rhetoric to bolster the booksellers' interests.

2   William Fisher divides copyright justifications into four categories: utilitarianism, labour theory, personality theory, and social planning theory. We have combined labour theory and personality theory together as "natural law" theories; what Fisher calls "social planning theory" is reflected in part in the section of this chapter on information as public good. Fisher, "Theories of Intellectual Property," 168–99.

3   Locke, *Second Treatise of Government*, 287–8.

4   Ibid., 290.

5   See Hughes, "Locke's 1694 Memorandum."

6   *Cobbett's Parliamentary History*, 1078 (1813). For a further analysis of the case see Rose, "The Author as Proprietor."

7   A similar result was reached in the United States in the 1834 case of *Wheaton v. Peters*, where the Supreme Court rejected the argument of common-law copyright in favour of a strict reading of the statute. For discussion of this case, see Patterson, *Copyright in Historical Perspective*, ch. 10; and McGill, "The Matter of the Text."

8 Civil law systems date back to Roman law, and are based on codes that set out the specific provisions of law. These codes are then applied and interpreted by judges when disputes arise. In contrast to code-based legal systems, the English common law is based on custom and practice as reflected in judicial precedent. For an accessible explanation of Canada's history of combining both "civil law" and "common law" systems, see Canada in the Making, www.canadiana.ca/citm/themes/constitution/constitution8_e.html.

9 Bentham, *Introduction to the Principles of Morals and Legislation*, ch. 1, article VII.

10 Section 91 of the Canadian Constitution merely lists copyright as an enumerated power of the federal government, with no rationale or guidance provided. See also Murray, "Protecting Ourselves to Death"; and *Théberge v. Galerie d'Art du Petit Champlain Inc.*, para. 32.

11 Henry Richardson argues that cost-benefit analysis "makes no room for intelligent deliberation about how to best use our resources," and that it thus "defeats its own aims." "The Stupidity of the Cost-Benefit Standard," 136.

12 Similarly, section 27(8) of the Patent Act states that "no patent shall be granted for any mere scientific principle or abstract theorem."

13 Examples are advance knowledge of weather conditions that will send the price of a crop's futures soaring, or inside financial data indicating that a company will have to restate its books to show a large loss. There are whole bodies of law protecting trade secrets and confidential information, and prohibiting certain uses of insider corporate information.

2. COPYRIGHT'S HISTORIES

1 For the text of the Statute of Anne, see Avalon Project, Yale University, http://avalon.law.yale.edu.

2 On the history of the print trade and incipient copyright in this period, see Rose, *Authors and Owners*, especially chs. 2, 3; Feather, "From Rights in Copies to Copyright," 191–209; and Loewenstein, *The Author's Due*.

3 Loewenstein, *The Author's Due*, 214.

4 Daniel Defoe, "Essay on the Regulation of the Press" (1704), quoted in Loewenstein, *The Author's Due*, 215.

5 Rose, *Authors and Owners*, 43. It is worth noting how the rhetoric of the Stationers' Company bears a similarity with that of the modern recording industry. In both cases changes in technology, the economy, and cultural practices challenge the settled order or, as we would say today, existing business models. See, for example, Hatch, "Toward a Principled Approach to Copyright Legislation at the Turn of the Millennium."

6 Patterson, *Copyright in Historical Perspective*, 148.

7 In a memorable phrase, the counsel for the appellant in *Donaldson v. Becket* (1774) said that booksellers "had not, till lately, ever concerned themselves about authors . . . nor would they probably have, of late years, introduced the authors as parties in their claims to the common law right of exclusively multiplying copies, had not they found it necessary to give a colourable face to their monopoly." Quoted in Loewenstein, *The Author's Due*, 14.

8 For differing accounts of *Donaldson v. Becket*, see Rose, *Authors and Owners*, ch. 6, and Loewenstein, *The Author's Due*, ch. 1. Johns retells this story in a striking new way ("Of Epics and Orreries," ch. 6 of *Piracy*, 109–43), identifying the contrast between literary writings and scientific inventions as a key feature of the debate.

9 Under this authority, Congress has enacted various laws including the Copyright Act, the Patent Act, and the Semi-Conductor and Chip Protection Act of 1984 (codified respectively at 17 USC sections 101, *et. seq.*; 35 USC sections 1, *et. seq.*; and 17 USC sections 901–14).

10 For examples of courts' invocation of this principle, see *Sony Corp. of America v. Universal City Studios, Inc.*; *Feist Publications v. Rural Telephone Service*.

11 For a discussion of the relationship between censorship and copyright law, see, for example, Cohen, "Constitutional Issues Involving Use of the Internet"; and Netanel, "Copyright and a Democratic Civil Society."

12 "Nothing in this act shall be construed to extend to prohibit the importation or vending, reprinting, or publishing within the United States, of any map, chart, book or books, written, printed, or published by any person not a citizen of the United States, in foreign parts or places without the jurisdiction of the United States." Copyright Act of 1790, stat. 124, section 5; www.copyright.gov/history/1790act.pdf.

13 McGill, *American Literature and the Culture of Reprinting*, 1.

14 The claim that copyright was the cause of the delay in American literary development is not well backed by evidence. See for example Khan, "Copyright Piracy and Development."

15 Justice Oliver Wendell Holmes, quoted in Hesse, "The Rise of Intellectual Property," 42. For more on *Bleistein v. Donaldson*, see Zimmerman, "The Story of Bleistein v. Donaldson Lithographic Company," 77–108.

16 Hesse, "The Rise of Intellectual Property," 42. For extensive commentary on the shift to expansionism in U.S. copyright law, see Vaidhyanathan, *Copyrights and Copywrongs*; Boyle, *Shamans, Software, and Spleens*; and Lessig, *Free Culture*.

17 Hesse, "The Rise of Intellectual Property," 42.

18 For information on the early history of French copyright, see Hesse, *Publishing and Cultural Politics in Revolutionary Paris*; and Ginsburg, "A Tale of Two Copyrights."

19 Davies, *Copyright and the Public Interest*, 152.

20 Copyright Act (France, 1957) quoted in Davies, *Copyright and the Public Interest*, 153.

21 As a matter of federal jurisdiction, copyright law is identical across Canada. However, francophone and anglophone judges come from different jurisprudential or intellectual traditions. For a suggestive discussion of the murky relationship between "copyright" and *"droit d'auteur"* in Canada, see Tawfik, "Copyright as Droit d'Auteur," 59–81.

22 WIPO (www.wipo.int) is a United Nations agency that administers international intellectual property agreements such as the Berne Convention. It is separate and apart from the WTO, which causes a considerable amount of confusion and concern that the proliferation of treaties is causing unnecessary overlap and complexity.

23 Parker, *The Beginnings of the Book Trade in Canada*, 109. See also Nadel, "Copyright, Empire and the Politics of Print," which acutely emphasizes the imperial dynamics of the situation.

24 Parker, *The Beginnings of the Book Trade in Canada*, 130. Eventually publishers in Britain began to realize that they simply could not sell to Canadians at British prices. In the 1850s the Commissioners of National Education in Ireland even gave away their copyrights on textbooks to British North American booksellers in an effort to address concerns about the negative influence of U.S. materials.

25 Parker, *The Beginnings of the Book Trade in Canada*, 168–69, 173.

26 John Lovell quoted in Parker, *The Beginnings of the Book Trade in Canada*, 174.

27 Parker, *The Beginnings of the Book Trade in Canada*, 193; and for a fuller account, MacLaren, *Dominion and Agency*, chs. 1 and 2.

28 MacLaren, *Dominion and Agency*, 13; and ch. 5.

29 MacLaren, *Dominion and Agency*, 14.

30 MacLaren, *Dominion & Agency*, 141–42.

31 The Act was amended in 1931, 1935, 1936, 1938, 1971, 1988, 1993, 1995, 1996, and 1997. This includes amendments as part of implementing particular international agreements. For example, the North American Free Trade Implementation Act of 1993 and the World Trade Organization Agreement Implementation Act of 1994 both contained provisions amending the Copyright Act. As for complaints about lack of action, John Kennedy, chairman and CEO of the International Federation of the Phonographic Industry (IFPI), was only one promulgator of the mantra: "It's astonishing that a sophisticated nation like Canada has dragged its feet for so long while the rest of the world has adapted its copyright laws to the digital age." CRIA (Canadian Recording Industry Association) press release, 2 March 2006.

32 These reports include the Isley Commission, 1957 (officially the Royal Commission on Patents, Copyright, Trademarks and Industrial Design); Report on Intellectual and Industrial Property, 1971 (Economic Council of Canada); Keynes and Burnett

Report, 1977 ("Copyright in Canada—Proposals for Revision of the Law"); Copyright Revision Studies, 1980–83 (fourteen separate studies on different aspects of copyright reform with emphasis on economic issues); "From Gutenberg to Telidon: A White Paper on Copyright: Proposals for the Revision of the Canadian Copyright Act," 1984; A Charter of Rights for Creators, 1985; Final Report, Copyright Subcommittee of the Working Group on Canadian Content and Culture: Information Highway Advisory Council (Industry Canada, 1995); "Connection, Community, Content: The Challenge of the Information Highway," Final Report of the Information Highway Advisory Council, 1995 (Industry Canada); Consultation Paper on Digital Copyright Issues, 2001 (Industry Canada and Canadian Heritage); Supporting Culture and Innovation: Report on the Provisions and Operation of the Copyright Act, 2002 (often referred to as the Section 92 Report); Status Report on Copyright Reform submitted to the Standing Committee on Canadian Heritage by the Minister of Canadian Heritage and the Minister of Industry, 24 March 2004; Standing Committee on Canadian Heritage, Interim Report on Copyright Reform, May 2004 (often referred to as the Bulte Report); Government Statement on Copyright Reform, March 2005 (Industry Canada and Canadian Heritage); and most recently the Government of Canada Copyright Consultation of 2009.

33 Angus has continued to take the position he articulated to the *Timmins Daily Press* in March 2005: "Downloading and peer-to-peer sharing of music files can be seen as a nightmare for recording artists losing out on potential royalties, but it's also a way for independent musicians to promote their talent . . . There is a royalty problem but I think it has to be put into perspective . . . There's been an incredible opportunity to get creative content and control it themselves." He has also promoted a levy on iPods (see ch. 19), trying to steer a course that balances creator and consumer interests.

34 "Harper Pledges Law to Combat Camcording in Cinemas," CBC News, 31 May 2007.

35 Other blogs fostered users' rights conversations as well, including those of Russell McOrmond (Digital Copyright Canada), Howard Knopf (Excess Copyright), and Laura Murray (faircopyright.ca). Another important users' rights player throughout this decade was David Fewer at CIPPIC, the Canadian Internet Policy and Public Interest Clinic at University of Ottawa.

36 The Appropriation Art Coalition was initiated by artist-curators Gordon Duggan and Sarah Joyce.

37 For a good discussion of how treaties and trade agreements affect Canadian copyright policy, see Tawfik, "International Copyright Law," 66–85.

3. COPYRIGHT'S SCOPE

1 For an overview of Canadian intellectual property laws, see Judge and Gervais, *Intellectual Property*, and Vaver, *Intellectual Property Law*. There are other intellectual property laws besides the "big four." In Canada, federal Acts that deal with special instances of intellectual property include the Industrial Design Act (1985), Plant Breeders' Rights Act (1990), and Integrated Circuit Topography Act (1990).

2 Just as the Canadian Supreme Court has recently affirmed the idea of balance in copyright law, it has done the same in patent. In *Teva Canada Ltd. v. Pfizer Canada Inc.*, the court said, "The patent system is based on a 'bargain', or quid pro quo: the inventor is granted exclusive rights in a new and useful invention for a limited period in exchange for disclosure of the invention so that society can benefit from this knowledge. This is the basic policy rationale underlying the Act. The patent bargain encourages innovation and advances science and technology" (para. 32). See also *Monsanto Canada Inc. v. Schmeiser* and *Harvard College v. Canada*. In the area of trademarks, see *Mattel, Inc. v. 3894207 Canada Inc.* and *Veuve Clicquot Ponsardin v. Boutiques Cliquot Ltée*.

3 Hayhurst, "Copyright Subject Matter," 31.

4 *Canadian Admiral Ltd. v. Rediffusion Inc.*, 91.

5 This determination was ultimately reversed by the Appeals Court and the Supreme Court of Canada.

6 *Tele-Direct v. American Business Information, Inc.*, para. 28. Leave to appeal to the Supreme Court was denied.

7 *CCH v. Law Society of Upper Canada*, para. 23. Later in the decision, the court grounded its placement of the originality standard in policy terms, stating: "When an author must exercise skill and judgment to ground originality in a work, there is a safeguard against the author being overcompensated for his or her work. This helps ensure that there is room for the public domain to flourish as others are able to produce new works by building on the ideas and information contained in the works of others."

8 The U.S. statute does spell out the fixation requirement. It grants protection to "original works of authorship fixed in any tangible medium of expression, now known or later developed, from which they can be perceived, reproduced, or otherwise communicated, either directly or with the aid of a machine or device" (section 102), and states that "a work is 'fixed' in a tangible medium of expression when its embodiment in a copy or phonorecord, by or under the authority of the author, is sufficiently permanent or stable to permit it to be perceived, reproduced, or otherwise communicated for a period of more than transitory duration" (section 101).

9 Vaver, *Copyright Law*, 65.

10 When Jana Sterbak's *Vanitas: Flesh Dress for an Albino Anorectic* (1987) was exhibited in Ottawa's National Gallery, it caused quite a controversy as it slowly decayed. For

more information, see http://art-history.concordia.ca/eea/artists/sterbak.html.

11 The right to the first fixation of a performance belongs to the performer, as section 15(1)(a)(iii) gives the performer the right to fix a performance in any material form where it is not yet fixed—but also grants copyright protection to the unfixed performance.

12 For the linguist Ferdinand de Saussure, language structures our very experience of reality, an idea elaborated by psychoanalyst Jacques Lacan, Marxist scholars of ideology, including Louis Althusser, and various anthropologists and feminist scholars. With different emphases, these and other scholars argue that language is not a label for underlying facts or ideas (or for an underlying "self"), but rather brings these facts or ideas into being. See Schleifer and Rupp, "Structuralism."

13 A similar conclusion was reached in the classic U.S. case *Baker v. Selden* (1879). In that case the court allowed copyright in a book describing accounting methods, but not in the methods themselves or the particular forms devised to facilitate them: "The very object of publishing a book on science or the useful arts," the court said, "is to communicate to the world the useful knowledge which it contains. But this object would be frustrated if the knowledge could not be used without incurring the guilt of piracy of the book." Quoted in Litman, "The Public Domain," 981.

14 *Anne of Green Gables Licensing Authority Inc. v. Avonlea Traditions Inc.*, para. 100, 121.

15 *Delrina Corp. v. Triolet Systems Inc.*, para. 52.

16 The lack of a formality requirement is a significant difference between copyright and patents. Patents are only issued after an application and formal examination process.

17 Copyright Act, section 39(2); see ch. 8 here.

18 Canadian Intellectual Property Office, "Guide to Copyrights," www.cipo.ic.gc.ca.

19 The dates for the transitional periods are calculated somewhat differently in Canadian Intellectual Property Office, "Guide to Copyrights." While an attempt was made in 2004 to extend the phase-out period for an unpublished work of an author who died before 1 January 1949, in what came to be known as the "Lucy Maud Montgomery Act" provisions of Bill C-8, this effort ultimately failed. See Banks and Hébert, "Legislative History of Bill C-8" and Howard Knopf, "Mouse in the House: A New Bill in Ottawa Appears to Adopt U.S.-Style Copyright Term Extension," *National Post*, 7 June 2003, FP 11.

4. OWNERS' RIGHTS

1 In contrast to the specific enumeration of separate instances of owners' rights in sections 3(1)(a) through (j), section 106(2) of the U.S. Copyright Act gives the owner the general right "to prepare derivative works based upon the copyrighted work."

2 Reproduction is also a highly problematic concept with regard to computer use. Drassinower, "Taking User Rights Seriously," 473, states: "Digital technology ruptures the continuity between copyright theory and copyright doctrine, such that the concept of reproduction no longer adequately separates infringing from noninfringing use. Applied in the digital environment, the right of reproduction grants owners the exclusive right to view their works where such viewing requires—as it does in the case of 'browsing'—the making of temporary copies. Thus, to insist on reproduction as the central organizing category of copyright law is to upset the copyright balance so as to grant owners a new and unprecedented control of access to copyrighted works."

3 Earlier Copyright Board cases also demonstrated the inadequacy of analysis based on quantitative factors alone. See "Licence Application by Pointe-a-Callière, Montreal Museum of Archeology and History for the Reproduction of Quotations," Copyright Board, 29 March 2004; "Re: Breakthrough Films," Copyright Board, 6 March 2006. *Warman v. Fournier*'s tests for substantiality have roots in *Hager v. ECW Press* (1999). An appeal in this case was filed by the *National Post* in March 2013. See Howard Knopf, "Warman v Fournier: Copyright in Titles and Headlines of Newspaper Articles?" http://excesscopyright.blogspot.ca, 14 March 2013.

4 Although broadcast law and technology have changed since this case, it remains appropriate for its thinking on what is meant by "public" performance.

5 Section 29.5(d) permits, on the premises of an educational institution for educational purposes, "the performance in public of a cinematographic work, as long as the work is not an infringing copy or the person responsible for the performance has no reasonable grounds to believe that it is an infringing copy."

6 The definition does get more complex, as section 2.2(1) provides:
For the purposes of this Act, "publication" means
(*a*) in relation to works,
   (i) making copies of a work available to the public,
   (ii) the construction of an architectural work, and
   (iii) the incorporation of an artistic work into an architectural work, and
(*b*) in relation to sound recordings, making copies of a sound recording available to the public,
but does not include
(*c*) the performance in public, or the communication to the public by telecommunication, of a literary, dramatic, musical or artistic work or a sound recording, or
(*d*) the exhibition in public of an artistic work.

7 The court was specifically dealing with the telecommunication right, in section 3(1) (f) as a subcategory of the public performance right, and it also indicated, by way of example, that the rental rights in section 3(1)(i) "can fit comfortably into the general

category of reproduction rights" (*ESA v. SOCAN*, para. 42). While the court did not go through each one of the separately listed rights in section 3(1)(a) through (i), it would appear that the translation right in section 3(1)(a) and the conversion and adaptation rights in sections (3)(1)(b) through (e) would also fit into the reproduction right. Section 3(1)(g), with respect to the public exhibition of artistic works, seems more analogous to the public performance right.

8　The majority of the Federal Court of Appeal agreed, and the Supreme Court of Canada declined to review the case. The case on appeal dealt with a host of issues; here we focus only on translation rights.

9　While there are no cases on this point, it would seem that conversion rights should also be applicable to new media, for example, in a situation where a work is "translated" into a multimedia presentation or a video game.

10　An analysis produced by the government explained the clause as follows: ". . . when a book has first been published, the author has the right to ensure that each copy of that book first enters the market through authorized dealers. Once ownership of that copy of the book has been transferred by the copyright owner, wherever in the world, the right is 'exhausted' or terminated with respect to that particular copy of the book." See Michael Geist, "Behind the Scenes of Bill C-32: Govt's Clause-by-Clause Analysis Raises Constitutional Questions," www.michaelgeist.ca, 27 September 2011.

11　The first sale doctrine is codified in section 109(a) of the U.S. Copyright Act. In *Kirtsaeng v. John Wiley* (2013), the U.S. Supreme Court gave it a broad construction, rejecting a publisher's claim that it did not apply to books lawfully purchased outside of the United States. See Howard Knopf, "Nifty Shades of Gray Marketing," http:// excesscopyright.blogspot.ca, 20 March 2013.

12　It seems odd that under new section 14.1(1), moral rights are offered only to those who perform "aurally" and record their work in sound recordings. It is not clear to us why silent performers would not have moral rights.

13　The Visual Artists Rights Act of 1990 added section 106A to the U.S. Copyright Act, setting forth certain moral rights for visual arts and adding a definition of visual arts to section 101: "to include (1) a painting, drawing, print or sculpture, existing in a single copy, in a limited edition of 200 copies or fewer that are signed and consecutively numbered by the author, or, in the case of a sculpture, in multiple cast, carved, or fabricated sculptures of 200 or fewer that are consecutively numbered by the author and bear the signature or other identifying mark of the author; or (2) a still photographic image produced for exhibition purposes only, existing in a single copy that is signed by the author, or in a limited edition of 200 copies or fewer that are signed and consecutively numbered by the author." Among the limitations to the definition are "works for hire."

5. USERS' RIGHTS

1 *Society of Composers, Authors and Music Publishers of Canada (SOCAN) v. Canadian Association of Internet Providers (CAIP)*, para. 40. For the specific importance of the *SOCAN* case with respect to the communication right, see ch. 4.

2 See, for example, *Basic Books, Inc. v. Kinko's Graphics Corp.*; *American Geophysical Union v. Texaco, Inc.*; *Princeton University Press v. Michigan Document Services*; *BMG Music v. Gonzalez*. For a summary of these and other U.S. fair use cases, see http://fairuse.stanford.edu.

3 Vaidhyanathan, "Copyright Law and Creativity."

4 *Cambridge University Press v. Becker* (known as the Georgia State case) and *Authors Guild v. HathiTrust* are discussed in ch. 16; see also Aufderheide and Jaszi, *Reclaiming Fair Use*, ch. 6. It could be said that a certain area of "tolerated use" may be opening up, in which rights owners may decide that the cost of bad publicity may make lawsuits too expensive. Thus many DJs and video mash-up artists have never been sued, even though they make no secret of their borrowings

5 At the time of the *CCH* decision, education, satire, and parody were not purposes listed in the Copyright Act; we have added them into the *CCH* tests.

6 See ch. 16 for a discussion of *Cambridge University Press v. Becker* and *Authors Guild v. HathiTrust*, in which U.S. courts distinguished between fictional and factual or educational works under the test for the "nature of the work."

7 Some exceptions for particular industries or bureaucratic necessities have been deemed outside the scope of this book. For example, broadcasters are permitted to make ephemeral copies of works, performances, and sound recordings as needed according to the logistics of the situation, with certain conditions (Copyright Act, sections 30.8, 30.9); section 32.1(1) allows copies or transmission of information for the purpose of complying with the Access to Information Act, Privacy Act, Broadcast Act, and Cultural Property Import and Export Act.

6. COLLECTIVES AND THE COPYRIGHT BOARD

1 For a complete list and description of the Canadian collectives, along with links to their websites, see the Copyright Board website, www.cb-cda.gc.ca/societies-societes/index-e.html.

2 For example, Copibec represents reproduction rights in works in Quebec and Access Copyright covers the rest of Canada.

3 For a listing of SOCAN's various tariffs, see www.socan.ca/licensees.

4 See Howard Knopf, "Working Committee on the Operations, Procedures and Processes of the Copyright Board," http://excesscopyright.blogspot.ca, 26 November 2012.

5 Interview with Laura Murray, 4 April 2012.

6 Cory Doctorow, remarks following keynote address at SIGGRAPH 2011, Vancouver (a conference on computer graphics).

7 The proposal in its January 2011 version can be found at www.songwriters.ca/proposalsummary.aspx.

8 D.C. Reid, "Writers—Focus on the Money," http://creatorsac.blogspot.ca, 18 September 2011. See also other posts on this website, many referring to Access Copyright's implementation or non-implementation of the Friedland Report, delivered to Access in February 2007 but not released for a year later. See Michael Geist, "Independent Report Blasts Access Copyright over Lack of Transparency," www.michaelgeist.ca, 15 February 2008, and Laura Murray, "A Cooperative Future for Access Copyright?" www.faircopyright.ca, 20 February 2008.

7. DETERMINING OWNERSHIP

1 The Reproduction of Federal Law Order (SI/97-5) states: "Anyone may, without charge or request for permission, reproduce enactments and consolidations of enactments of the Government of Canada, and decisions and reasons for decisions of federally-constituted courts and administrative tribunals, provided due diligence is exercised in ensuring the accuracy of the materials reproduced and the reproduction is not represented as an official version." Similar orders also pertain to most of the provinces.

2 For a thorough analysis of arguments for abolishing Crown copyright, see Judge, "Crown Copyright and Copyright Reform in Canada." For the United States, S.C. section 105 provides: "Copyright protection under this title is not available for any work of the United States Government, but the United States Government is not precluded from receiving and holding copyrights transferred to it by assignment, bequest, or otherwise."

3 See Copyright Board of Canada website, www.cb-cda.gc.ca. DeBeer and Bouchard, "Canada's 'Orphan Works' Regime," provides a detailed study of how this system is working.

4 See U.S. Copyright Office, "The Importance of Orphan Works Legislation," www.copyright.gov, 25 September 2008; and the Association of Research Libraries, "Resource Packet on Orphan Works," www.arl.org, September 2011.

8. enforcement of owners' rights

1 In *Milliken & Co. v. Interface Flooring Systems (Canada) Inc.* the court said that while the burden of showing knowledge rests with the plaintiff, it can be inferred from the facts. In this case a designer had been retained by the defendant to design a carpet tile on a large project. Even though the designer was not an employee or officer of the corporation, the knowledge was imputed to the defendant because she was given charge of the project, and based on her expertise in the field the court found she should have known she was infringing.

2 Quoted in Michael Geist, "Why Liability Is Limited: A Primer on New Copyright Damages as File Sharing Lawsuits Head to Canada," www.michaelgeist.ca, 28 November 2012.

3 In the United States there is a further reduction of damages provision that could in essence be adopted in Canada. Under section 504 of the U.S. Copyright Act, the court is directed to remit damages in cases where the infringer believed and had reasonable grounds for believing that the copyrighted work was a fair use under section 107 (17 USC 107), if the infringer was working for a non-profit educational institution, library, or archives (or under certain circumstances a public broadcasting entity). This section provides employees of institutions that are trying to help members of the public a certain degree of protection against damage awards where their assessment of fair use was wrong but made in good faith. While the Canadian courts have the discretion to reduce the award in such a circumstance, they do not have to.

4 Many complaints have been collected on the website ExtortionLetterInfo (www.extortionletterinfo.com) under the keyword "Masterfile"; for advice for those receiving such letters, see the CIPPIC website at www.cippic.ca/Trolls.

5 Howard Knopf, "A Cautionary Tale of Costly Copyright Litigation Consequences: How to Win a Little and Lose a Lot," http://excesscopyright.blogspot.ca, 18 November 2012.

9. music

1 The most recent round of this familiar controversy was the showdown between The American Assembly's 2012 report "Media Piracy in Emerging Economies" (http://piracy.americanassembly.org), and studies by NPD, a firm hired by the Recording Industry of America. See Joe Karaganis, "Pirates Are the Best Customers: Anatomy of a Ridiculous Controversy," http://infojustice.org, 19 November 2012.

2 Section 2 of the Act defines "musical work" as "any work of music or musical composition, with or without words, and includes any compilation thereof."

3 "Sound recording maker" is defined in the Copyright Act as "the person by whom the arrangements necessary for the first fixation of the sounds are undertaken," with the elaboration that such arrangements may "include arrangements for entering into contracts with performers, financial arrangements and technical arrangements." Note that the maker is not the engineer, the person at the soundboard, but rather the person who makes the recording happen. In many cases, the "maker" may be the performer; in other situations, it is a label. Like performers' rights, makers' rights are limited to those associated with Rome Convention countries; see Copyright Act section 18(2).

4 In the case of a performer's performance, the term is "50 years after the end of the calendar year in which the performance occurs" (section 23[1]), but if the performance is subsequently fixed in a sound recording before the end of the term it will extend for another 50 years from the end of the calendar year in which the first fixation occurs (section 23[1][a]). The term for sound recordings is "50 years after the end of the calendar year in which the first fixation of the sound recording occurs" and if the sound recording is published within the term it extends for another 50 years from the end of the calendar year of publication (section 23[1.1]).

5 For information on specific collectives and tariffs, see the Copyright Board of Canada website, www.cb-cda.gc.ca.

6 Since 1997 Canadian law provides performers and sound-recording makers from Rome Convention countries with an equitable right to remuneration with composers. The International Convention for the Protection of Performers, Producers of Phonograms, and Broadcasting Organizations was negotiated at Rome in 1961; it is administered by the World Intellectual Property Organization. For the details of Rome's conditions and members, see Copyright Act 15(2) and www.wipo.int/treaties/en/ip/rome. Canada is a party to this treaty, but the United States is not, so performers there do not hold rights except by individual contract.

7 Section 17.1 and 17.2.

8 For the full text and commentary on the leading U.S. copyright infringement cases dealing with music, see UCLA's Music Copyright Infringement Resource, http://mcir.usc.edu. While U.S. precedents do not necessarily hold in Canada, the recording industry is dominated by U.S.-based labels, so the normal practice has been founded in U.S. case law. It is unclear, however, whether cases like *Grand Upright v. Warner* and *Bridgeport Music* would be applied in Canada.

9 Jane Taber, "Tories Go Negative with iPod-Tax Coalition Fear Campaign," *Globe and Mail*, 16 December 2010; for a more measured response, see Michael Geist, "Angus Introducing Private Copying Levy Bill, Flexible Fair Dealing Motion," www.michaelgeist.ca, 16 March 2010.

10 "Where Do Music Collections Come From?" Media Piracy in Emerging Economies, http://piracy.americanassembly.org, 15 October 2012.

11 Joel Waldfogel, "Bye, Bye, Miss American Pie? The Supply of New Recorded Music since Napster," NBER Working Paper No. w16882, March 2011, http://papers.ssrn.com.

12 Luc VNO, comment on Michael Geist, "U.S. Paper Says No Decline in New Music in Napster Age," www.michaelgeist.ca, 23 March 2011.

10. DIGITAL MEDIA

1 If you put a notice explicitly requiring permission for educational use, educational users will not be able to use the PAM (Publicly Available Material) exception (section 30.04) and will either have to rely on fair dealing or seek your permission for use.

2 See http://creativecommons.ca.

3 *Globe and Mail*, Terms and Conditions (para 3), www.theglobeandmail.com/help/terms-and-conditions.

4 www.cbc.ca/permissions/faq-general.html.

5 For example, the company Readex specializes in digitizing public domain sources; see www.newsbank.com/readex. While Readex is not reducing the public's access to the original documents—most are available in special collections at research libraries—it is preserving a certain inequity of access, because the price for subscription is steep. A different approach is represented by public access digitization projects such as Project Gutenberg (www.gutenberg.org), HathiTrust (www.hathitrust.org and see ch. 16), and the International Music Score Library Project (http://imslp.org and see ch. 3).

6 The right of resale is codified in the United States as the "first sale doctrine." While it is not codified in Canadian law, it is equally established in custom and commerce. See Reese, "The First Sale Doctrine in the Era of Digital Networks," 577.

7 When you rip open the shrinkwrap that encloses a software package, or click through an agreement online, you may be giving your assent to an enforceable contract. It's important, therefore, to read and understand the terms. However, Canadian law on this topic is still emerging and not always consistent. See Morgan, "I Click, You Click, We all Click"; Sigel et al., "Validity of Webwrap Contracts."

8 www.exxonmobil.com/siteflow/Notices/SF_MS_LegalNotice_TC.aspx

9 www.nyx.com/terms-use (see under Use of Links)

10 http://en.wikipedia.org/wiki/Deep_linking

11 See Michael Geist, "Canadian Federal Court Says No Copyright Infringement for Linking," www.michaelgeist.ca, 25 June 2012; Samuel Trosow, "A Closer Look at Warman v Fournier: Good News for Journalists," http://samtrosow.wordpress.com, 26 June 2012.

12 See Gervais, "User-Generated Content and Music File-Sharing"; McKenzie et al, "User-Generated Online Content 1"; and McNally et al, "User-Generated Online Content 2."

13 Given the text of the conditions in section 29.21 (a) through (d), the UGC exception appears to have several parallels to fair dealing. But while it is closely related, it is not exactly the same. The two defences complement each other and someone defending against an infringement action could raise either or both exceptions. In a situation where one of the threshold categories of fair dealing is not present, the defence under section 29.21 is still available assuming all of its conditions can be met. On the other hand, where a use is commercial or where a potential exploitation could have an adverse effect, fair dealing is not necessarily nullified—it would depend on all of the factors—whereas this exception would be.

14 Halbert, "Mass Culture and the Culture of the Masses," 930.

11. FILM, VIDEO, AND PHOTOGRAPHY

1 See *Mattel, Inc. v. 3894207 Canada Inc.*, and *Veuve Clicquot Ponsardin v. Boutiques Cliquot Ltée.*

2 Sara Perry, "Fight or Flight!" www.entertainmentmedialawsignal.com, 30 November 2012.

3 A photographer commissioned by the Quebec tourist board to take photographs with famous Quebec landmarks as backdrops eventually abandoned his attempts to feature the 1969 Jean-Paul Riopelle sculpture *La Joute* near Montreal's Palais de Congrès. After the Riopelle estate claimed the right to authorize such a photograph, the photographer, wanting to avoid a large fee or a lawsuit, walked away. His actions were understandable given the headaches of going to court, but the Riopelle estate probably had no case. See Baillargeon, "Hauteurs et bassesses du droit d'auteur."

4 Kevin McMahon, interview with Laura Murray, Toronto, 16 November 2006.

5 "Canada's Documentary Film Heritage and Its Future in Jeopardy," press release, 4 December 2006, www.docorg.ca.

6 "The Search for Fair Dealing: Report on the DOC Road Show," 19 October 2011; available at www.docorg.ca.

7 Documentary Organization of Canada, "Fair Dealing and Copyright: Guidelines for Documentary Filmmakers," May 2010; Documentary Filmmakers' Statement of Best Practices in Fair Use, available through the Center for Social Media, November 2005.

8 Walter Forsberg, remark in discussion at Copycamp, an "Unconference" sponsored by the Creators' Rights Alliance and others at Ryerson University, Toronto, 30 September 2006.

I 2. VISUAL ARTS

1 Canadian Artists' Representation (CARFAC), "Court Sides with National Gallery on Artists' Fees," press release, 8 March 2013; *National Gallery of Canada v. Canadian Artists' Representation*.

2 A letter to Ministers of Canadian Heritage and Industry, June 2006, Appropriation Art website.

3 Smith, "When One Man's Video Art Is Another's Copyright Crime."

4 *Suntrust v. Houghton Mifflin*. Randall did not in fact use any character names or words from Mitchell's novel—but whether or not she would have been infringing outside of fair use was never resolved because the case was settled out of court. It might also be noted that as Margaret Mitchell died in 1949, *Gone with the Wind* has been in the Canadian public domain since 2000, so there was never a copyright issue in Canada; in the United States *Gone with the Wind* will not enter the public domain until 2020.

5 *Rogers v. Koons*. See also Jaszi, "On the Author Effect," 29–56.

6 "Inuit Artists to Miss Out on Resale Right Payments at Coming Auctions," www.carfac.ca, 1 November 2012.

I 3. CRAFT AND DESIGN

1 To find out if a particular industrial design is registered, see the database of registered industrial designs in the Canadian Intellectual Property Office's online "Guide to Industrial Designs."

2 See Storey, "Filing Design Applications in Canada and the United States," 123–49.

3 U.S. Copyright Act, 17 U.S.C. § 101.

4 In the United States, there has been considerable discussion about the desirability of offering copyright-like protection to fashion design, which has resulted so far in the passage of the Innovative Design Protection Act through the Senate Judiciary Committee in September 2012. Critics of the idea have argued that the current legal situation may actually benefit the commercial producers that it appears to leave "unprotected." As Kal Raustiala puts it in "Fashion Victims": "Copying drives the fashion cycle. Unlike areas such as software or cell-phones [in which] the more people who use your software or the more callers on your network, the better off you are, fashion designs become progressively less attractive as they saturate a market.

... The hunger for design distinctiveness drives fashion lovers back into Barneys, Bergdorfs, and the boutiques on a regular (and ever-quickening) basis."

See also Raustiala and Sprigman, *The Knockoff Economy*. For a comparative introduction to the Canadian situation, see Courtney Doagoo, "Third Time a Charm? The Innovative Design Protection Act in the Face of The Knockoff Economy," www.iposgoode.ca, 12 November 2012.

5 Public shaming has been used widely rather than legal avenues: for example, an entire blog, Urban Counterfeiters, was set up to track and trace the apparent perfidy of Urban Outfitters, a company accused of stealing the designs of independent creators.

6 See Halbert, "The Labor of Creativity"; Kathleen Bissett, "Copyright: How It Affects the Quilter," www.canadianquilter.com; and "Quilt Kerfuffle Blankets Competitions," *Eastern Ontario AgriNews*, www.agrinewsinteractive.com, March 2006.

7 Interview with Laura Murray, Kingston Ontario, 15 April 2011.

8 Robertson, "Craft and Copyright."

14. JOURNALISM

1 See Vaver, *Copyright Law*, 83–90. Employees might argue that some work done on their own time or initiative belongs to them.

2 "Where the work is an article or other contribution to a newspaper, magazine or similar periodical," section 13(3) of the Copyright Act reserves "to the author a right to restrain the publication of the work, otherwise than as part of a newspaper, magazine or similar periodical." Historically, this might have been something like a moral right, protecting the writer's reputation from unwanted associations. But *Robertson v. Thomson* suggests that the right is economically valuable in an environment in which publishers can make money from republishing in electronic formats—then again, this confirmation of its value probably means that employers will be even more likely to clear it away by contract.

3 See Vaver, *Copyright Law*, 84–86, 89–90, 95–97.

4 Among the many irate responses to this contract in 2004, see Brian Brennan, "Sign Here, in Blood," http://brianbrennan.ca/blog, 4 December 2004.

5 Canadian Professional Writers Survey, May 2006, available at www.pwac.ca/about/pwacadvocacy.

6 Professional Writers Association of Canada, www.pwac.ca/about/faqanswers2.

7 *Robertson v. Thomson*, 70.

8 This decision may also hamper digitization of small periodicals—the arts magazines and local papers that form the cutting edge or grassroots of Canadian culture. These

publications don't have the staff or budget to track down past contributors. But perhaps it is mainly a matter of choosing the right software: the *Robertson v. Thomson* case suggests that if a publisher emphasizes the integrity of the publication as a whole, clearance from authors is not necessary. Or perhaps a royalty or revenue-sharing arrangement for digital uses unimagined at the time of original publication could work, along with special grants for small magazines and small publishers to migrate their materials to electronic archives. Fair dealing would be another way forward, given the expansive language of the 2012 Supreme Court cases. The challenge will be to find a mechanism that will fairly and productively accommodate not only the megapublishers and their contributors, but also the small publishers and their contributors, while allowing preservation of Canada's literary and journalistic heritage.

9 *Toronto Star*, Pages of the Past, http://thestar.pagesofthepast.ca.

10 British Library, "Intellectual Property: A Balance. The British Library Manifesto," 25 September 2006.

11 Brad Wheeler, "Look, It's Trying to Pop out of My Bra Right Now!" *Globe and Mail*, 23 October 2006, R1.

15. EDUCATION

1 Chris Tabor, Director, Queen's Campus Bookstore, email, 10 December 2012.

2 Jane Phillips, Coordinator of Collection Development, Queen's University Library, email, 10 December 2012.

3 Some educational institutions are also concerned with the possibility that should they not sign a licence with Access Copyright, they could be liable for back fees for such a licence, should one case of infringement be found at their institution. We do not find this risk to be substantial: for one thing, the penalty would only be the expense avoided in the first place, and for another, a court would not likely interpret section 68.2(1) so harshly against the interests of educational institutions.

4 See Association of Research Libraries, "Code of Best Practices in Fair Use," www.arl.org/focus-areas/copyright-ip/fair-use/code-of-best-practices, January 2012.

5 *Alberta v. Access Copyright*, para. 35. It should be noted that the dissent agreed with the majority on this last point regarding the unreasonable nature of the Board's conclusion on this factor (para. 57).

6 The Act also features a few other exceptions that may be used by educational institutions along with religious, charitable, and fraternal organizations (see Table 8, pp. 84–86).

7 Section 2 of the Copyright Act defines "educational institution" as
    (*a*) a non-profit institution licensed or recognized by or under an Act of

Parliament or the legislature of a province to provide pre-school, elementary, secondary or post-secondary education,

(b) a non-profit institution that is directed or controlled by a board of education regulated by or under an Act of the legislature of a province and that provides continuing, professional or vocational education or training,

(c) a department or agency of any order of government, or any nonprofit body, that controls or supervises education or training referred to in paragraph (a) or (b), or

(d) any other non-profit institution prescribed by regulation

8   Section 2 states: "'premises' means, in relation to an educational institution, a place where education or training referred to in the definition 'educational institution' is provided, controlled or supervised by the educational institution." This definition creates difficulties for institutions that are trying to conduct outreach beyond their physical campuses, for instance through distance education or a community outreach program.

9   Section 29.4(3) now reads: "Except in the case of manual reproduction, the exemption from copyright infringement provided by subsections (1) and (2) does not apply if the work or other subject-matter is commercially available, within the meaning of paragraph (a) of the definition 'commercially available' in section 2, in a medium that is appropriate for the purposes referred to in those subsections."

10 Many educational institutions have had licences with the collectives Criterion and Audio Ciné to cover the "public performance" of film materials. Given the addition of films to section 29.5, it is doubtful that these licences are still needed.

11 "For the purposes of this section, 'lesson' means a lesson, test or examination, or part of one, in which, or during the course of which, an act is done in respect of a work or other subject-matter by an educational institution or a person acting under its authority that would otherwise be an infringement of copyright but is permitted under a limitation or exception under this Act."

12 As the President of Athabasca University put it: "Students are expected to somehow accumulate knowledge as they proceed through their studies. The content delivered in one course builds on the knowledge acquired in previous courses. The provision that content from Algebra 1 must be destroyed so that students taking Algebra 2 cannot refer back to it when needed is counter to the principles of education and how people learn. It just does not make sense." Frits Pannekoek, Letter Concerning Proposed Copyright Changes, 18 November 2008.

13 Section 30.02 allows certain digital copying and communication of works subject to several counter-limitations and conditions. It only applies if the institution has a licence with a collective society that permits making reprographic copies (i.e., Access Copyright) and the institution must take measures to prevent students from making more than one copy or further reproducing or communicating that copy.

Section 30.03 requires that if a subsequent licence or tariff has a different rate, then the difference must be paid. These exceptionally complex provisions seem intended to lock in an unnecessary dependency on licences. See Trosow, "Bill C-32 and the Educational Sector," 560–2.

14 The performance and communication exceptions are limited to where the public is primarily students or other persons under the institution's authority.

15 See sections 30.04(3), with respect to an access control, and 30.04(4)(a), with respect to a use control. See ch. 8 for a discussion of what constitutes such a "technological protection measure."

16 Section 30.04(4)(b). The section does not explain what constitutes such a clearly visible notice; that will be developed through regulation.

17 Supporters included the Association of Universities and Colleges of Canada (AUCC) (see News Release, "Proposed Copyright Law Amendments: Some Very Good Changes but Some Cause for Concern," 13 June 2008); Council of Ministers of Education of Canada (CMEC) (see news release, "CMEC Copyright Consortium Pleased with New Federal Copyright Legislation," 13 June 2008); Canadian Association of University Libraries (CARL) (see Brent Roe, Submission to 2009 Copyright Consultations). Objectors included Canadian Association of University Teachers (CAUT) (see Copyright and Academic Staff, *CAUT Education Review* 1.10 [Feb. 2008]) and the Canadian Federation of Students (CFS) (see "Statement on Copyright Reform," 2008). The Canadian Library Association (CLA) also expressed objections; see "Unlocking the Public Interest: The Views of the Canadian Library Association on Bill C-61, An Act to Amend the Copyright Act," September 2008.

18 Murray, "Protecting Ourselves to Death"; Sam Trosow, "Educational Use of the Internet Amendment: Is it Necessary?" http://samtrosow.ca, 31 January 2008; Howard Knopf, "The 'A Contrario' Scenario & CMEC," http://excesscopyright.blogspot.ca, 31 January 2008, and "The CMEC Red Herring," http://excesscopyright.blogspot.ca, 13 March 2008.

19 Trosow, "Bill C-32 and the Educational Sector," 545–51.

20 Broadly speaking, Part A of the Access Copyright licences allowed multiple copies for classroom or administrative use of "up to 10% of a published work" or the following, whichever is greater:
an entire chapter that is less than 20 per cent of the book's length
an entire article or page from a periodical
an entire story, poem, essay, etc. from a book or periodical containing other such pieces
an entire artistic item from a book containing other such items
an entire encyclopedia or dictionary item
There is a developing consensus in the educational community, as reflected in a number of emerging fair dealing guidelines, that these uses are all within fair dealing as

it has been articulated in *Alberta v. Access Copyright* (2012). See Michael Geist, "Educational Fair Dealing Policy Shows Why the Access Copyright License Provides Little Value," www.michaelgeist.ca, 1 October 2012.

21 Statement of Proposed Royalties to Be Collected by Access Copyright for the Reprographic Reproduction, in Canada, of Works in Its Repertoire (*Canada Gazette* 144.24 [12 June 2010]). A "licence" is a voluntary agreement entered into by the parties. If the parties cannot agree on a licence, then they may apply to the Copyright Board for a "tariff" (see ch. 6). In April 2013, Access Copyright filed a new post-secondary education tariff with the Copyright Board, this one for the period 2014–17. It contains the same controversial provisions carried over from the 2011–13 application. (Access Copyright, "Canada's Writers and Publishers Take a Stand against Damaging Interpretations of Fair Dealing by the Education Sector," press release, www.accesscopyright.ca, 8 April 2013.

22 See the joint objection filed by the Canadian Federation of Students and the Canadian Association of University Teachers, 11 August 2010. For a full discussion of the grounds for objection to the tariff, see Trosow, Armstrong, and Harasym, "Objections to the Proposed Access Copyright Post-Secondary Tariff and Its Progeny Licenses," 2012.

23 See Sam Trosow, "Toronto and Western Sign Licensing Agreement with Access Copyright," http://samtrosow.wordpress.com, 31 January 2012; Howard Knopf, "UofT and Western Capitulate to Access Copyright," http://excesscopyright.blogspot.ca, 31 January 2012; and Canadian Association of University Teachers (CAUT), "Copyright Agreement with Western and Toronto a Bad and Unwarranted Deal," www.caut.ca, 2 February 2012.

24 AUCC did not post its model licence on its website, but see CAUT, "A Bad Deal: AUCC/Access Copyright Model License Agreement," www.caut.ca, n.d.

25 For a list of schools signing or rejecting the agreement, see Ariel Katz, "Fair Dealing's Hall of F/Sh/ame," http://arielkatz.org, 15 May 2012; and Sam Trosow, "Compilation of Announcements for Institutions Opting-out of Model License," http://samtrosow.wordpress.com, 29 June 2012.

26 See Michael Geist, "ACCC Legal Counsel: Access Copyright Licence Provides 'Little Value,'" www.michaelgeist.ca, 6 September 2012. Geist cites a legal opinion from ACCC's counsel concluding that "[t]here is therefore little value in remaining in a tariff relationship with Access Copyright beyond August 31, 2012." Similarly, the Ontario Association of School Boards has advised its members to stop using the Access Copyright tariff. Michael Geist, "Ontario Public School Boards Preparing to Drop Access Copyright Next Year," www.michaelgeist.ca, 15 October 2012.

27 Access Copyright, "Canada's Writers and Publishers Take a Stand against Damaging Interpretations of Fair Dealing by the Education Sector," press release, www.accesscopyright.ca, 8 April 2013.

16. LIBRARIES, ARCHIVES, AND MUSEUMS

1   According to section 2 of the Copyright Act, "'library, archive or museum' means (*a*) an institution, whether or not incorporated, that is not established or conducted for profit or that does not form a part of, or is not administered or directly or indirectly controlled by, a body that is established or conducted for profit, in which is held and maintained a collection of documents and other materials that is open to the public or to researchers, or (*b*) any other non-profit institution prescribed by regulation."

2   While to date there is no Canadian case law that raises the issue of transformativity, it could easily be a dimension of the first or second fairness tests from the *CCH* case—the purpose or the character of the use—and thus may yet have an explicit place in Canadian law.

3   Section 30.1(2) provides that paragraphs 30.1(1)(a) to (c) do not apply where an appropriate copy is commercially available in a medium and of a quality that is appropriate for the purposes of section 30.1(1).

4   According to the American Library Association, ILL service "is intended to complement local collections and is not a substitute for good library collections" and "is based on a tradition of sharing resources between various types and sizes of libraries." Interlibrary Loan Code for the United States Explanatory Supplement (section 2), on www.ala.org. See also the Canadian University Reciprocal Borrowing Agreement, www.curba.ca.

5   American Library Association, "Fair Use and Electronic Resources," www.ala.org/advocacy/copyright/fairuse/fairuseandelectronicreserves.

6   *CCH v. Law Society of Upper Canada*, 2004, para. 61.

7   Bielstein, *Permissions*, 101.

8   Zorich, "Developing Intellectual Property Policies." Of course, insofar as collections include works under copyright, would-be users must set the price of reproduction with the actual copyright owner, and museums have to negotiate terms for online exhibition. See ch. 12 for related issues. Our focus here is on the public domain, the easiest place for LAMs to start with a renewed strategy for cultural public service.

9   Panitch and Michalak, "The Serials Crisis."

10 For a Directory of Open Access Journals, see www.doaj.org. For examples of Open Access journals, see the Public Library of Science, www.plos.org; and BioMed Central, www.biomedcentral.com. See also SPARC Open Access Newsletter (maintained by Peter Suber), www.earlham.edu/~peters/fos.

11 See Canadian Association of Research Libraries, "E-Books in Research Libraries"; Horava, "E-books Licensing and Canadian Copyright Legislation."

17. COPYRIGHT'S COUNTERPARTS

1 See Macdonald, *Lessons of Everyday Law*.

2 See Fauchart and Von Hippel, "Norms-Based Intellectual Property Systems"; Oliar and Sprigman, "There's No Free Laugh (Anymore)"; Loshin, "Secrets Revealed"; Fagundes, "Talk Derby to Me."

3 For resistance to pharmaceutical and biotech commodification, see Indigenous Peoples Council on Biocolonialism website, www.ipcb.org. The World Intellectual Property Organization has also been a focus of international activism through its Intergovernmental Committee on Intellectual Property and Genetic Resources, Traditional Knowledge and Folklore (www.wipo.int/tk). See also Coombe, *The Cultural Life of Intellectual Properties*; Bell and Napoleon, *First Nations Cultural Heritage and Law*; and for a range of discussion of cultural appropriation, Ziff and Rao, eds., *Borrowed Power*.

4 Caroline Anawak, for example, writes: "Indigenous artists must take great care not to interpret what they are given. They must be true to the bond and deliberately refrain from changing that which is not changeable" ("Indigenous Knowledge and Artistic Expression").

5 This is, of course, a generalization. There are as many distinct customary laws as there are First Nations, Indigenous people in Canada have been subjected to Canadian law for several generations, cultures have changed, and Indigenous individuals live and work in the city and participate in the art world or the marketplace. Tension sometimes arises over the authority to make decisions about use of cultural traditions. But there is a broad consensus in Indigenous circles about the idea of responsibility to the history and future of the people, and a widespread frustration and anger at non-Indigenous appropriation of Indigenous culture.

6 Bringhurst quoted in entry for Ghandl, BC BookWorld, www.abcbookworld.com.

7 Jusquan, "Yew Wood to Bringhurst: A Story of Indigenous Knowledge."

8 Because of the independent but overlapping nature of citation and copyright economies, you can infringe copyright without plagiarizing (if you cite your source but you don't ask permission), and you can plagiarize without infringing copyright (if, for example, you take from a source in the public domain). See Murray, "Plagiarism and Copyright Infringement."

9 Bakhtin, *The Dialogic Imagination*, 293.

10 This is the name of the leading academic "citation index," a way in which influence can be quantified by the number of times a given paper has been cited. Thus the rather romantic metaphor of a "web" is turned into a "productivity indicator": our point being that citation systems are not necessarily "nicer" than market systems—just different.

11 The Ministry of Canadian Heritage has sponsored various studies on the subject of educating children about copyright. Access Copyright unveiled Captain Copyright, an education resource, in 2006, but withdrew the site's content the same year after a wave of criticism at the biased nature of the information.

12 Quoted by Susan Crean, Copycamp, Toronto, September 2006. Also see Lorinc, "Creators and Copyright in Canada."

13 http://en.wikipedia.org/wiki/Open_source, accessed 11 December 2012.

14 See www.ubuntu.com and http://opensource.org.

15 Johan Söderberg states: "The distinguishing and most promising feature of free software is that it has mushroomed spontaneously and entirely outside of previous capital structures of production. It has built a parallel economy that outperforms the market economy" ("Copyleft vs. Copyright").

16 Creative Commons Canada, "'Some Rights Reserved.'"

17 See Murray and Robertson, "Appropriation Appropriated."

18 Documentary Organization of Canada, "Canada's Documentary Film Heritage and Its Future in Jeopardy," press release, 5 December 2006, www.docorg.ca. On museum copyright policies, see ch. 16.

19 Holden, "Cultural Value and the Crisis of Legitimacy."

20 See http://plr-dpp.ca, and for a timeline that indicates a wide range Canadian cultural policy ideas and initiatives over the years, Duxbury et al, "Cultural Infrastructure."

21 Michael Schrage, "Why Giveaways Are Changing the Rules of Business," *Financial Times*, 6 February 2006.

18. COPYRIGHT'S FUTURE

1 Available at www.parl.gc.ca/legisinfo. See Howard Knopf, "Bill C-56: Just When You Thought It Was Safe to Go Back into the Water?" http://excesscopyright.blogspot.ca, 4 March 2013; and Michael Geist, "What's Really Behind Canada's Anti-Counterfeiting Bill?" www.michaelgeist.ca, 13 March 2013.

2 See Michael Geist, "Access Copyright's Desperate Declaration of War against Fair Dealing," www.michaelgeist.ca, 9 April 2013; and Howard Knopf, "Access Copyright Thrashes Thrice," http://excesscopyright.blogspot.ca, 8 April 2013.

3 Access Copyright, "Canada's Writers and Publishers Take a Stand against Damaging Interpretations of Fair Dealing by the Education Sector," press release, www.accesscopyright.ca, 8 April 2013.

Legal Citations and Cases

Here is a list of all the cases we have mentioned in this book. As intimidating as it may seem, you may find that making sense of legal rulings is not that difficult. They are refreshingly easier to understand than the Copyright Act itself, and often very interesting to read and argue with.

With most Canadian cases, we are using simplified citations from the Canadian Legal Information Institute (CanLII). CanLII is part of an international "open law" movement including sites in various other countries. Its website (www.canlii.org) does not require a subscription and is open to the public. Some downsides: it is not (yet) complete, and not all the older cases have page or paragraph numbering, so it can be hard to find precise parts of the judgments. But because of its accessibility, we have used CanLII citations where they are available. In CanLII, *CCH v. Law Society of Upper Canada* is cited as 2004 SCC 13, to indicate it is the thirteenth case from the Supreme Court of Canada issued in 2004. CanLII uses abbreviations for the level of the court including FC for Federal Court and ON CA for Ontario Court of Appeal.

Canadian Supreme Court cases are also available online from Lexum at the University of Montreal, http://scc.lexum.org. The U.S. Legal Information Institute, which contains a much lower proportion of that country's court cases, is run out of Cornell Law School: www.law.cornell.edu. We have used traditional legal citation format for all cases not available on LII sites.

Legal Cases Cited

Alberta (Education) v. Access Copyright, 2012 SCC 37 (CanLII).

Allen v. Toronto Star Newspapers Ltd. (1997) 36 O.R. (3d) 201(Div.Court).

American Geophysical Union v. Texaco, Inc., 60 F.3d 913 (2d Cir. 1995).

Anne of Green Gables Licensing Authority v. Avonlea Traditions Inc., 2000 CanLII 5698 (ON CA).

Apple Computer v. Mackintosh Computer Ltd. 10, C.P.R. (3d) 1 (Fed Trial Court, 1986); [1988] 1 F.C. 673; [1990] 2 S.C.R. 209.

Authors Guild, Inc. v. HathiTrust, ___ F. Supp. 2d ____. 2012 U.S. Dist. LEXIS 146169 (S.D.N.Y. Oct. 10, 2012) (appeal filed 8 November 2012).

Baker v. Selden, 101 U.S. 99 (1879).

Basic Books, Inc. v. Kinko's Graphics Corp., 758 F.Supp. 1522 (S.D.N.Y. 1991).

B.C. Jockey Club v. Standen, 1985 CanLII 591 (BC CA).

Bleistein v. Donaldson, 188 U.S. 239 (1903).

BMG Canada v. Doe, 2004 FC 488; 2005 FCA 193 (CanLII).

BMG Music v. Gonzalez, 430 F.3d 888 (7th Cir. 2005).

Bridgeport Music v. Dimension Films, 410 F. 3d 792 (6th Cir. 2005).

Cambridge University Press v. Becker (Georgia State University), 863 F. Supp. 2d 1190 (N.D. Ga 2012).

Campbell v. Acuff-Rose Music (1994) 510 U.S. 596.

Canadian Admiral Ltd. v. Rediffusion Inc. [1954] Exchequer Court Rep. 382, 20 C.P.R. 75.

CCH Canadian Ltd. v. Law Society of Upper Canada, 1999 CanLII 7479 (FC); 2002 FCA 187 (CanLII); 2004 SCC 13 (CanLII).

Compagnie Générale des Établissements Michelin-Michelin & Cie v. National Automobile, Aerospace, Transportation and General Workers Union of Canada (CAW-Canada), 1996 CanLII 3920 (FC).

Cuisenaire v. South West Imports Ltd., 1968 CanLII 122 (SCC).

Delrina Corp. v. Triolet Systems Inc., 2002 CanLII 11389 (ON CA).

Dr. Bonham's Case, 77 Eng. Rep. 638 (Common Pleas,1610).

Donaldson v. Becket, 98 Eng. Rep., 4 Burr. 2408 (H.L. 174).

Entertainment Software Association (ESA) v. Society of Composers, Authors and Music Publishers of Canada (SOCAN), 2012 SCC 34 (CanLII).

Feist Publications v. Rural Telephone Service, 499 U.S. 340 (1991).

Georgia State. *See* Cambridge University Press.

Glenn Gould Estate v. Stoddart Publishing Co. Ltd., 1998 CanLII 5513 (ON CA).

Grand Upright v. Warner, 780 F. Supp. 182 (S.D.N.Y.) (1991).

Hager v. ECW Press Ltd., 1998 CanLII 9115 (FC).

Harvard College v. Canada (Commissioner of Patents), 2002 SCC 76 (CanLII).

Kirtsaeng v. John Wiley & Sons, Inc., 568 U.S. ___ , Supreme Court 11–697 (decided 19 March 2013).

Mattel, Inc. v. 3894207 Canada Inc., 2006 SCC 22 (CanLII).

Michelin v. CAW. See Compagnie Générale des Établissements Michelin.

Milliken & Co. v. Interface Flooring Systems (Canada) Inc., 1998 CanLII 9044 (FC) and 2000 CanLII 14871 (FCA).

Monsanto Canada Inc. v. Schmeiser, 2004 SCC 34 (CanLII).

National Gallery of Canada v. Canadian Artists' Representation, 2013 FCA 64 (CanLII).

Princeton University Press v. Michigan Document Services, 99 F.3d 1381 (6th Cir. 1996), cert. denied 520 U.S. 1156 (1997).

Prise de Parole Inc. v. Guérin (1995) 66 C.P.R. (3d) 257 (Federal Court Trial Div), aff'd (1996) 73 C.P.R. (3d) 557 (Fed. C.A.).

Prism Hospital Software v. Hospital Medical Records Institute, 57 C.P.R. (3d) 129 (B.C.S.C., 1994).

Re:Sound v. Motion Picture Theatre Associations of Canada, 2012 SCC 38 (CanLII).

Robertson v. Thomson Corp., 2006 SCC 43 (CanLII).

Rogers v. Koons, 751 F. Supp. 474 (S.D.N.Y. 1990), upheld at 960 F.2d 301 (2d Cir. 1992).

Rogers Communications Inc. v. Society of Composers, Authors and Music Publishers of Canada (SOCAN), 2012 SCC 35 (CanLII).

Society of Composers, Authors and Music Publishers of Canada (SOCAN) v. Bell Canada, 2012 SCC 36 (CanLII).

Society of Composers, Authors and Music Publishers of Canada (SOCAN) v. Canadian Association of Internet Providers (CAIP), 1 C.P.R. (4th) 417 (Copyright Board, 1999); 2002 FCA 166 (CanLII); 2004 SCC 45 (CanLII).

Snow v. Eaton Centre Ltd., 70 C.P.R. (2d) 105 (Ontario High Court of Justice).

Sony Corp. of America v. Universal City Studios, Inc., 464 U.S. 417 (1984).

Stowe v. Thomas, 23 F. Cas. 201 (C.C.E.D. Pa. 1853) (No. 13,514).

Suntrust Bank v. Houghton Mifflin, 268 F.3d 1267 (11th Cir. 2001).

Tele-Direct (Publications) Inc. v. American Business Information, Inc., 1997 CanLII 6378 (FCA).

Teva Canada Ltd. v. Pfizer Canada Inc., 2012 SCC 60 (CanLII).

Théberge v. Galerie d'Art du Petit Champlain Inc., 2002 SCC 34 (CanLII).

Veuve Clicquot Ponsardin v. Boutiques Cliquot Ltée, 2006 SCC 23 (CanLII).

Warman v. Fournier, 2012 FC 803 (CanLII).

Wheaton v. Peters (1834), 33 US 591.

Bibliography

Anawak, Caroline. "Indigenous Knowledge and Artistic Expression." Discussion paper prepared for National Gatherings on Indigenous Knowledge, Department of Canadian Heritage, 2005.

Aufderheide, Patricia, and Peter Jaszi. *Reclaiming Fair Use: How to Put Balance Back in Copyright*. Chicago: University of Chicago Press, 2011.

Baillargeon, Stéphane. "Hauteurs et bassesses du droit d'auteur." *Le Devoir*, 31 August 2006.

Bakhtin, Mikhail. *The Dialogic Imagination*. Ed. Michael Holquist. Trans. Caryl Emerson and Michael Holquist. Austin: University of Texas Press, 1981.

Banks, Sam, and Monique Hébert. "Legislative History of Bill C-8: The Library and Archives of Canada Act." Parliament of Canada, Law and Government Division, 18 February 2004.

Bell, Catherine, and Val Napoleon. *First Nations Cultural Heritage and Law: Case Studies, Voices, and Perspectives*. Vancouver: University of British Columbia Press, 2008.

Benkler, Yochai. "From Consumers to Users: Shifting the Deeper Structures of Regulation towards Sustainable Commons and User Access." *Federal Communications Law Journal* 52.3 (May 2000).

Bentham, Jeremy. *Introduction to the Principles of Morals and Legislation*. 1780.

Bielstein, Susan M. *Permissions, A Survival Guide: Blunt Talk about Art as Intellectual Property*. Chicago: University of Chicago Press, 2006.

Boateng, Boatema. *The Copyright Thing Doesn't Work Here: Adinkra and Kente Cloth and Intellectual Property in Ghana*. Minneapolis: University of Minnesota Press, 2011.

Boswell, James. *Boswell's Life of Johnson*. 1791. London: Oxford University Press, 1961.

Boyle, James. *Shamans, Software, and Spleens: Law and the Construction of the Information Society*. Cambridge, Mass.: Harvard University Press, 1996.

———. "The Second Enclosure Movement and the Construction of the Public Domain." *Law and Contemporary Problems* 66 (2003).

Canadian Association of Research Libraries Copyright Committee Task Group on E-Books. "E-Books in Research Libraries: Issues of Access and Use." April 2008.

Cohen, Julie E. "Constitutional Issues Involving Use of the Internet: Intellectual Property and Censorship of the Internet." *Seton Hall Constitutional Law Journal* 8 (1998).

Coombe, Rosemary J. *The Cultural Life of Intellectual Properties: Authorship, Appropriation, and the Law*. Durham, N.C.: Duke University Press, 1998.

Creative Commons Canada. "'Some Rights Reserved': Building a Layer of Reasonable Copyright." http://creativecommons.ca.

Davies, Gillian. *Copyright and the Public Interest*. London: Sweet and Maxwell, 2002.

DeBeer, Jeremy F., and Mario Bouchard. "Canada's 'Orphan Works' Regime: Unlocatable Owners and the Copyright Board." *Oxford University Commonwealth Law Journal* 10.2 (Winter 2010).

Drassinower, Abraham. "Taking User Rights Seriously." In *In the Public Interest*, ed. Geist.

Duxbury, Nancy, Christina Johnson, Kelsey Johnson, and Erin Schultz. "Cultural Infrastructure: A Chronology of Key Developments and Contexts." Appendix G to *Under Construction: The State of Cultural Infrastructure in Canada*. Vancouver: Centre of Expertise on Culture and Communities, Simon Fraser University, July 2008.

Fauchart, Emmanuelle, and Eric Von Hippel. "Norms-Based Intellectual Property Systems: The Case of French Chefs." *Organization Science* 19.2 (2008): 187–201.

Fagundes, David. "Talk Derby to Me: Emergent Intellectual Property Norms Governing Roller Derby Pseudonyms." *Texas Law Review* 90.5 (2012): 1093–1152.

Feather, John. "From Rights in Copies to Copyright: The Recognition of Authors' Rights in English Law and Practice in the Sixteenth and Seventeenth Centuries." In *The Construction of Authorship: Textual Appropriation in Law and Literature*, ed. Martha Woodmansee and Peter Jaszi. Durham, N.C.: Duke University Press, 1994.

Fisher, William W. "Theories of Intellectual Property." In *New Essays in the Legal and Political Theory of Property*, ed. Stephen R. Munzer. Cambridge and New York: Cambridge University Press, 2001.

Frye, Northrop. *Anatomy of Criticism*. Princeton, N.J.: Princeton University Press, 1957.

Geist, Michael, ed. *In the Public Interest: The Future of Canadian Copyright Law*. Toronto: Irwin Law, 2005.

———. *From "Radical Extremism" to "Balanced Copyright": Canadian Copyright and the Digital Agenda*. Toronto: Irwin Law, 2008.

Gervais, Daniel. "User-Generated Content and Music File-Sharing: A Look at Some of the More Interesting Aspects of Bill C-32." In *From "Radical Extremism" to "Balanced Copyright,"* ed. Geist.

Ginsburg, Jane C. "A Tale of Two Copyrights: Literary Property in Revolutionary France and America." *Tulane Law Review* 64 (1990).

Gravenor, Kristian. "Shot without Consent." *Montreal Mirror*, 4 August 2005.

Halbert, Debora J. "The Labor of Creativity: Women's Work, Quilting, and the Uncommodified Life." *Transformative Works and Cultures* 3 (2009).

———, "Mass Culture and the Culture of the Masses: A Manifesto for User-Generated Rights." *Vanderbilt Journal of Entertainment and Technology Law* 11 (2009).

Hatch, Orrin. "Toward a Principled Approach to Copyright Legislation at the Turn of the Millennium." *University of Pittsburgh Law Review* 59 (1998).

Hayhurst, William L. "Copyright Subject Matter." In *Copyright and Confidential Information Law in Canada*, ed. Gordon Henderson, Howard Knopf, and John Rudolph. Toronto: Carswell, 1994.

Hesse, Carla. *Publishing and Cultural Politics in Revolutionary Paris, 1789–1810*. Berkeley: University of California Press, 1991.

———. "The Rise of Intellectual Property, 700 B.C.–A.D. 2000: An Idea in the Balance." *Daedalus* 131 (2002).

Holden, John. "Cultural Value and the Crisis of Legitimacy: Why Culture Needs a Democratic Mandate." *Demos*, 29 March 2006.

Hughes, Justin. "Locke's 1694 Memorandum (and More Incomplete Copyright Historiographies)." Cardozo Legal Studies Research Paper 167, October 2006.

Horava, Tony. "E-books Licensing and Canadian Copyright Legislation: A Few Considerations." *Partnership: The Canadian Journal of Library and Information Practice and Research* 4.1 (2009).

Jaszi, Peter. "On the Author Effect: Contemporary Copyright and Collective Creativity." In *The Construction of Authorship: Textual Appropriation in Law and Literature*, ed. Martha Woodmansee and Peter Jaszi. Durham, N.C.: Duke University Press, 1994.

Jefferson, Thomas. *The Writings of Thomas Jefferson*. Constitution Society website, www.constitution.org.

Johns, Adrian. *Piracy: The Intellectual Property Wars from Gutenberg to Gates*. Chicago: University of Chicago Press, 2009.

Judge, Elizabeth F. "Crown Copyright and Copyright Reform in Canada." In *In the Public Interest*, ed. Geist.

Judge, Elizabeth F., and Daniel Gervais. *Intellectual Property: The Law in Canada*, 2nd ed. Toronto: Thomson-Carswell, 2011.

Jusquan, "Yew Wood to Bringhurst: A Story of Indigenous Knowledge." *Redwire*, Spring 2005. Reproduced in Koebel, "Aboriginal Youth and Traditional Knowledge: Honouring the Past and Acknowledging the Future." Discussion paper prepared for the National Gatherings on Indigenous Knowledge, Department of Canadian Heritage, 2005.

Khan, Zorina. "Copyright Piracy and Development: United States Evidence in the Nineteenth Century." *Revista de Economía Institucional*, 2008.

Lessig, Lawrence. *Free Culture: How Big Media Uses Technology and the Law to Lock Down Culture and Control Creativity*. London: Penguin, 2004.

Litman, Jessica. "The Public Domain." *Emory Law Journal* 39 (1990).

Locke, John. *Second Treatise of Government*. 1690. In *Two Treatises of Government*, ed. Peter Laslett. Cambridge: Cambridge University Press, 1960.

Loewenstein, Joseph. *The Author's Due: Printing and the Prehistory of Copyright*. Chicago: University of Chicago Press, 2002.

Lorinc, John. "Creators and Copyright in Canada." Prepared for the Creators' Copyright Coalition. www.creatorscopyright.ca/documents/lorinc.html, November 2004.

Loshin, Jacob. "Secrets Revealed: Protecting Magicians' Intellectual Property without Law." In *Law and Magic: A Collection of Essays*, ed. Christine A. Corcos, 123–42. Durham: Carolina Academic Press, 2010.

Macdonald, Roderick Alexander. *Lessons of Everyday Law*. Law Commission of Canada and School of Policy Studies, Queen's University. Kingston and Montreal: McGill-Queen's University Press, 2002.

MacLaren, Eli. *Dominion and Agency: Copyright and the Structuring of the Canadian Book Trade, 1867–1918*. Toronto: University of Toronto Press, 2011.

McGill, Meredith. "The Matter of the Text: Commerce, Print Culture, and the Authority of the State in American Copyright Law." *American Literary History* 9.1 (1997).

———. *American Literature and the Culture of Reprinting, 1834–1853*. Philadelphia: University of Pennsylvania Press, 2003.

McKenzie, Pamela J., Jacqueline Burkell, Lola Wong, Caroline Whippey, Michael McNally, and Samuel E. Trosow. "User-Generated Online Content 1: Overview, Current State and Context." *First Monday* 17.6 (June 2012).

McKeon, Michael. *The Secret History of Domesticity: Public, Private, and the Division of Knowledge*. Baltimore: John Hopkins University Press, 2005.

McNally, Michael, Samuel E. Trosow, Lola Wong, Caroline Whippey, Jacquelyn Burkell, and Pamela J. McKenzie. "User-Generated Online Content 2: Policy Implications." *First Monday* 17.6 (June 2012).

Morgan, Charles. "I Click, You Click, We All Click ... But Do We Have a Contract? A Case Comment on Aspencerl.com v. Paysystems." *Canadian Journal of Law and Technology* 14.2 (July 2005).

Mount, Nicholas J. *When Canadian Literature Moved to New York.* Toronto: University of Toronto Press, 2005.

Murray, Laura J. "Protecting Ourselves to Death: Canada, Copyright, and the Internet." *First Monday* 9.10 (2004).

———. "Plagiarism and Copyright Infringement: The Costs of Confusion." In *Originality, Imitation, and Plagiarism: Teaching Writing in the Digital Age,* ed. Caroline Eisner and Martha Vicinus. Ann Arbor: University of Michigan Press, 2008: 173–82.

Murray, Laura J., and Kirsty M. Robertson. "Appropriation Appropriated: Ethical, Artistic, and Legal Debates in Canada." In *Intellectual Property for the 21st Century: Multidisciplinary Approaches to Intellectual Property Law,* ed. Teresa Scassa, Mistrale Goudreau, Madelaine Saginur, and B. Courtney Doagoo. Toronto: Irwin Law, forthcoming.

Nadel, Ira B. "Copyright, Empire and the Politics of Print: The Case of Canada." Unpublished manuscript, 2006.

Netanel, Neil Weinstock. "Copyright and a Democratic Civil Society." *Yale Law Journal* 106 (1996).

Oliar, Dotan, and Christopher Sprigman. "There's No Free Laugh (Anymore): The Emergence of Intellectual Property Norms and the Transformation of Stand-Up Comedy." *Virginia Law Review* 94.8 (2008): 1787–1867.

Panitch, Judith M., and Sarah Michalak. "The Serials Crisis: A White Paper for the UNC–Chapel Hill Scholarly Communications Convocation." University of North Carolina, January 2005.

Parker, George L. *The Beginnings of the Book Trade in Canada.* Toronto: University of Toronto Press, 1985.

Patterson, Lyman Ray. *Copyright in Historical Perspective.* Nashville: Vanderbilt University Press, 1968.

Raustiala, Kal. "Fashion Victims," *The New Republic Online,* www.newrepublic.com, 15 March 2005.

Raustiala, Kal, and Chris Sprigman. *The Knockoff Economy: How Imitation Sparks Innovation.* NY: Oxford University Press, 2012.

Reese, R. Anthony. "The First Sale Doctrine in the Era of Digital Networks." *Boston College Law Review* 44 (March 2003).

Richardson, Henry S. "The Stupidity of the Cost-Benefit Standard." In *Cost Benefit Analysis: Legal, Economic, and Philosophical Perspectives*, ed. Matthew D. Adler and Eric A. Posner. Chicago: University of Chicago Press, 2001.

Robertson, Kirsty. "Craft and Copyright." Unpublished manuscript, December 2006.

Rose, Mark. "The Author as Proprietor: Donaldson v. Becket and the Genealogy of Modern Authorship." *Representations* 23 (Summer 1988).

———. *Authors and Owners: The Invention of Copyright*. Cambridge: Harvard University Press, 1993.

Saint-Amour, Paul K. *The Copywrights: Intellectual Property and the Literary Imagination*. Ithaca, N.Y.: Cornell University Press, 2003.

Schleifer, Ronald, and Gabriel Rupp. "Structuralism." In *The Johns Hopkins Guide to Literary Theory and Criticism*, 2nd ed., ed. Michael Groden and Martin Kreiswirth. Baltimore: Johns Hopkins University Press, 1994.

Sigel, Skip, Theo Ling, and Joshua Izenberg. "The Validity of Webwrap Contracts." Paper prepared for the Uniform Law Conference of Canada, 1999.

Smith, Roberta. "When One Man's Video Art Is Another's Copyright Crime." *New York Times*, 6 May 2004, E1, E5.

Söderberg, Johan. "Copyleft vs. Copyright: A Marxist Critique." *First Monday* 7.3 (4 March 2002).

Storey, R.B. "Filing Design Applications in Canada and the United States." *University of Baltimore Intellectual Property Law Journal* 10.2 (2002).

Tawfik, Myra J. "Copyright as Droit d'Auteur." *Intellectual Property Journal* 17 (2003).

———. "International Copyright Law: W[h]ither User Rights." In *In the Public Interest*, ed. Geist.

Trosow, Samuel E. "Bill C-32 and the Educational Sector: Overcoming Impediments to Fair Dealing." In *From "Radical Extremism" to "Balanced Copyright,"* ed. Geist.

Trosow, Samuel E., Scott Armstrong, and Brent Harasym. "Objections to the Proposed Access Copyright Post-secondary Tariff and Its Progeny Licenses: A Working Paper." http://ir.lib.uwo.ca/fimspub/24, 14 August 2012.

Trosow, Samuel E., Jacquelyn Burkell, Nick Dyer-Witheford, Pamela McKenzie, Michael B. McNally, Caroline Whippey, and Lola Wong. "Mobilizing User-Generated Content for Canada's Digital Advantage." Final Report for SSHRC Knowledge Synthesis Grant, http://ir.lib.uwo.ca/fimspub/21, 1 December 2010.

Vaidhyanathan, Siva. *Copyrights and Copywrongs: The Rise of Intellectual Property and How It Threatens Creativity*. New York: New York University Press, 2001.

———. "Copyright Law and Creativity." Presentation at Originality, Imitation, and Plagiarism, University of Michigan conference, 23–25 September 2005.

Vaver, David. *Intellectual Property Law: Copyright, Patents, Trade-Marks*. Toronto: Irwin Law, 2011.

———. *Copyright Law*. Toronto: Irwin Law, 2000.

Wiebe, Christle. "Customary Law and Cultural and Intellectual Properties." Discussion paper for Canadian Heritage Department Traditions Gatherings, 2005.

Young, Edward. "Conjectures on Original Composition." London: Millar and Dodsley, 1759.

Ziff, Bruce, and Pratima V. Rao, eds. *Borrowed Power: Essays on Cultural Appropriation*. New Brunswick, NJ: Rutgers University Press, 1997.

Zimmerman, Diane Leenheer. "The Story of Bleistein v. Donaldson Lithographic Company: Originality as a Vehicle for Copyright Inclusivity." In *Intellectual Property Stories*, ed. Jane C. Ginsburg and Rochelle Cooper Dreyfuss. New York: Foundation Press, 2006.

Zorich, Diane M. "Developing Intellectual Property Policies: A How-to Guide for Museums." Canadian Heritage Information Network, Professional Exchange, www.pro.rcip-chin.gc.ca.

Index

Abella, Justice Rosalie, 136, 176, 185–86, 218–19
academic journals: Open Access, 229; subscriptions to, 213–14
academic research, 225–26
access controls, 114, 257n15
Access Copyright, 81–82, 88, 90–91, 234–35, 247n2(ch.6), 248n8, 255n3; "Captain Copyright" website, 261n11; education and, 183, 185, 187; licensing and, 194–95, 197, 210
accessibility, universal, 232, 233
Access to Information Act, 247n7
Access to Knowledge (A2K), 228
activism: consumer rights, 30; users' rights, 29
adaptation, 63–64, 146; rights of, 245n7
add-ons, 147
Adobe Acrobat, 137
advocacy, 201; users' rights, 89; in visual arts, 158
agency, 95
Alberta (Education) v. Access Copyright (2012), 73, 78, 81–82, 151, 163, 183, 185–88, 200–201, 207, 208, 218
alienability, 221
Allen, Jan, 76
Allen, Jim, 181
Allen v. Toronto Star (1997), 180–81
Althusser, Louis, 244n12
American Assembly, 130; "Media Piracy in Emerging Economies," 249n1(ch.9)

American Declaration of Independence, 4
American Geophysical Union v. Texaco, 208, 247n2(ch.5)
American Library Association (ALA), 207, 259n4
Android systems, 147
Anne of Green Gables Licensing Authority v. Avonlea Traditions (2000), 46
Angus, Charlie, 30, 130, 242n33
anonymous publication, 224
anti-circumvention, 204–5
Anti-Counterfeiting Trade Agreement (ACTA), 32, 237
Apple, 129
Apple Computer v. Mackintosh Computer (1986), 63
appropriation, 162–64; from commons, 5; of physical resources, 5; of public resources, 4–5; undermining by, 8
appropriation art, 162–64; social benefits of, 162–63
Appropriation Art coalition, 31, 162, 242n36
architectural works, 38, 166
archives, 156–57, 199–214; digitization of, 236; management and maintenance of collections of, 203–5; special collections in, 211
archiving, 116, 176, 194
arrangers, 122, 124
art. *See* visual art
art co-operatives, Inuit, 220

artisans, law and, 167–70
artistic production, 4; value and precariousness of, 233
artistic works, 37, 38; adaptation of, 63–64; sole rights in, 56
artists: collective bargaining for, 160; as custodians and animators of collective tradition, 3; as re-creators, 4
art law clinics, 111
assigned rights, 97–99; partial, 97–98
association, right of, 68
Association of American Publishers, 196
Association of Community Colleges of Canada (ACCC), 195, 197
Association of Universities and Colleges of Canada (AUCC), 194–95
L'Atelier National du Manitoba, 154
attribution, 224; right of, 68, 126, 168
Audio Ciné, 87, 194, 256n10
Australia, Indigenous cultural materials in, 228
authorization, rights of, 66–67
Authors Guild, 201, 235
Authors Guild, Inc. v. HathiTrust (2011), 201–3, 207, 209, 235, 247n4, 247n6
authorship: individualistic and Romantic notion of, 43; joint, 224
authors' rights, 6–7, 19, 21, 22, 67–70, 238n1; limitations on, 94

backup copies, 83, 84, 116, 204–5
Baker v. Selden (1879), 244n13
Bakhtin, Mikhail, 225
balancing of interests, 10
Barbie's, 150–51
Basic Books, Inc. v. Kinko's Graphics Corp. (1991), 208, 247n2(ch.5)
B.C. Jockey Club v. Standen (1986), 40
Belyea, Susan, 170
Bemelmans, Ludwig, 52
Benkler, Yochai, 229
Bentham, Jeremy, 7
Berne Convention for the Protection of Literary and Artistic Works, 21, 25, 27–28, 31–32, 97
Between the Lines, 231
Beveridge, Karl, 159
Bieber, Justin, 229

Bielstein, Susan: Permissions: A Survival Guide, 211
Bill C-8, 244n19
Bill C-11 (Copyright Modernization Act, 2012), 28, 29, 31, 61, 113, 130, 145, 183, 191, 193, 197, 204–6, 218, 234, 237
Bill C-32 (2010), 28, 31
Bill C-56 (Combating Counterfeit Products Act), 234, 237
Bill C-60 (2005), 28
Bill C-61 (2008), 28, 30
Bissett, Kathleen, 168
blank media, levy on, 91, 130–31
Bleistein v. Donaldson, 21
blogosphere, as citation economy, 225
BMG Canada v. Doe, 128
BMG Music v. Gonzalez, 247n2(ch.5)
Boateng, Boatema: The Copyright Thing Doesn't Work Here, 219
body art, 44
books, 38; and DRM, 141–42; electronic, 142
booksellers: perpetual monopoly of, 19; rights of, 19
book trade: British, 18–19; Canadian, 26–28; French, 22; U.S., 20
Boswell, James: Boswell's Life of Johnson, 8
Boutique Cliquot, 150–51
Boyle, James: "The Second Enclosure Movement," 9
Boyle, John B., 226
Brett, Brian, 90
Bridgeport Music v. Dimension Films (2005), 127
Bringhurst, Robert, 222
British Columbia, 196
British Library, 178
British North America Act (1867), 24
Broadcast Act, 247n7
broadcasting: of films, 153; public, 232; public support for, 230
broadcast logistics, 115
broadcast rights, 125
broadcasts: copyright term for, 50; performance of in educational institutions, 189, 190–91
broadcast signals, 37, 98. See also communication signals
buildings and public art exception, 152
Bulte, Sarmite, 30

Bulte Report (Department of Canadian Heritage Interim Report on Copyright Reform, 2004), 30

cable television, 14
California Open Education Resources Council, 196
California Open Source Digital Library, 196
call-home mechanisms, 142–43
Cambridge University Press, 208
Cambridge University Press v. Becker, 207–9, 247n4, 247n6
camera technology, 149
Campbell v. Acuff-Rose Music (1994), 163
Canada Council, 233
Canadian Admiral v. Rediffusion (1954), 42–43, 61
Canadian Alliance of Dance Artists, 104
Canadian Artists' Representation Copyright Collective (CARCC), 64
Canadian Artists' Representation/Le Front des artistes canadiens (CARFAC), 160, 164
Canadian Association of Internet Providers (CAIP), 64, 65
Canadian Association of University Teachers, 258n22
Canadian Bar Association, 110
Canadian Broadcasting Corporation (CBC), 109–10, 138, 154, 157, 178, 232
Canadian Constitution, repatriation of, 27
Canadian Federation of Students, 258n22
Canadian Intellectual Property Office (CIPO), 48, 166–67
Canadian Internet Policy and Public Interest Clinic (CIPPIC), 107, 242n35
Canadian Library Association (CLA), 257n17
Canadian Musical Reproduction Rights Agency (CMRRA), 88, 126–27
Canadian Music Creators Coalition (CMCC), 130
Canadian Private Copying Collective (CPCC), 91
Canadian Recording Industry Association (CRIA), 29
Canwest/Global, 173
Canwest Mediaworks Inc., 177
Carroll, Jock, 179

cartoons, Internet, 140
cataloguing, 204
CCH v. Law Society of Upper Canada, 29, 40, 41, 60, 67, 72–74, 76, 100, 151, 155, 163, 164, 180, 209; and fair dealing in education, 183–85, 195; and libraries, 200–201, 207, 210–11; and six fairness factors for fair dealing, 78–82, 151, 184–85, 259n2
CCH Canadian Ltd., 72
CDs. *See* compact discs
censorship, 21; vs. copyright law, 240n11
Center for Social Media, Fair Use site of, 155
Central Canada Exhibition, 168
choreographic works, 38
Christian Copyright Licensing Inc. (CCLI), 87
cinematographic works. *See* films
circumvention, of digital locks, 29, 113–17, 204–5
citation, importance of, 227
citation economies, 224–26; why they work, 225
civil infringement, 101–10; primary, 102–3, 105; secondary, 102, 103–5
civil law, 6; vs. common law, 239n8
CKY (Winnipeg), 154
classroom displays, 188, 189
classroom screenings, 61
clearance: culture of, 155; high cost of, 156
Clement, Tony, 31
collaboration, , as domain of UGC, 147
collage, 68, 146, 162, 163
collections, management and maintenance of, 203–5
collective action, in visual arts, 158
collective bargaining, for artists, 160
collective creation, 224
collective licensing, 131, 183
collectives, 87–92, 93, 99, 235, 237; controversial nature of, 90–92; licensing arrangements administered by, 83; and music, 124–25
Combating Counterfeit Products Act. *See* Bill C-56
commercial cinema, 157
common law, 5; vs. civil law, 239n8
common sense, 218–19
communication rights, 199
communication signals, 39, 43; first ownership of, 95; rights in, 55. *See also* broadcast signals

community, Indigenous ownership and, 222
community colleges, 188, 193, 195, 235–36
compact discs (CDs), 122; backups of, 205;
digital copies of, 57
compilations, 38
composers, 122, 124, 125
computer code, 63
computer programs, 38, 63, 66; backing up, 83,
84; interoperability of, 83, 85, 115; and paracopyright, 142; sole rights in, 56
Condorcet, Marquis de, 22
confidential information, 35, 36
consent, 134–38; implied, 135, 137; implied vs.
explicit, 103; website, 134–38
Conservative Party, 30–31
consortial purchasing, 214
consumer rights, 30
consumption, rivalry in, 11–14, 141
content provider vs. user, 134
contracts, importance of, 104
conversation, as citation economy, 225
conversion of works, rights of, 63–64, 245n7
Copibec, 247n2(ch.6)
Copps, Sheila, 181
copy controls, 114
copyleft, 226, 228, 230
copyright: alternatives to, 219–30; balance
in, 71, 73; and "big four" of intellectual
property, 35–36; counterparts of, 217–33;
Crown, 96, 237; as cultural policy mechanism, 219; determining ownership of, 93–100;
as disappointment or irritant, 217; duration
of, 49–53, 98, 100, 124, 126; economic
analysis of, 8–10; established properties of,
4–10; exceptions to, 151–53, 184, 187–94,
203–4; exclusive interests, 10; as extension of
personality of author, 7; and facts and ideas,
45–47; formalities not required in, 47–49; as
form of private property, 49; future of, 234–
37; histories of, 16–32; history of in Canada,
16–17, 24–28; Indigenous statements on,
221; infringement of, 30, 47, 62, 68, 71–74,
79, 83–86, 101–17, 125, 128, 137, 141, 142,
144–45, 151–52, 161–62, 164, 171, 172, 177,
181, 183, 188, 191, 201, 205, 208–10, 218,
224; lack of central registry of, 93; legacies
of in Canada, 17–24; Lockean view of, 5; as

market mechanism, 219; meaning of, 56;
misuse of, 237; modern Canadian, 28–32; as
monopoly, 55, 219; as natural right, 5; and
original expression fixed in tangible form,
40–47; perpetual, 5–6; philosophies of, 4–10;
protections, expansion of, 10; and public
domain, 49–53; rationales for, 3–15; reform
of, 28–31; registration of, 47–49, 97; restrictions vs. protections, 10; scope of, 35–53;
student work, 193; subsisting in works and
other subject matter, 37–40; as system of
relationships and interests, 73; teachers'
work, 193; term of protection for, 49–53;
unlocatable ownership of, 99–100; vs. droit
d'auteur, 241n21
Copyright Act, 37, 45, 51, 54–56, 60, 61, 101,
138, 143, 151, 172, 174, 183, 194, 228, 234;
(1872), 25; (1875), 25; (1889), 27; (1924),
24, 27, 28; amendments to, 51, 187, 241n31;
Part IV, 113; Part VII, 87; review of (2017),
237; sections of: s2: 123, 126, 133, 245n6,
254n2, 255n7, 256n8, 259n1; s3: 55–67, 69,
95, 123, 159; s6: 50; s9: 50; s12: 50, 95; s13:
94–95, 173; s14: 67–69, 246n12; s15: 95,
123–24, 250n6; s17, 250n7; s18: 95, 126,
250n3; s21: 95; s23: 50, 250n4; s24: 95; s26:
95, 179; s27: 102, 103, 105; s28: 67–69, 160–
61; s29: 73, 83–86, 127, 140, 145–46, 148,
152–53, 164, 184–85, 188–91, 204–5, 245n5,
252n13, 256n10; s30: 151–52, 190–93, 203,
205–6, 211, 247n7, 251n1, 259n3; s32: 152,
247n7; s34: 105; ss34–41: 102; s35: 105; s38:
105, 107–8; s39: 105, 108; s41: 113–16; s42:
110, 112; ss42–43: 102; s43: 112; s64: 166,
171; s70: 88; s80: 128; ss81–86: 91; technological neutrality principle, 136
Copyright Board, 87–92, 93, 99, 136, 140, 187,
195, 197, 219, 235, 245n3; Interim Tariff,
235
copyright collective societies. See collectives
"copyright creep," 43
copyright interest: existence of at moment of
fixation, 47–49; limited duration of, 49–53
Copyright Modernization Act. See Bill C-11
copyright reform, 28–31; reports on, 242n32
copyright symbol, 47–48
corporate logos, 150–51

corporate ownership, 224
The Corporation, 155
cost-benefit analysis, 10
cost recovery, in LAMs, 211–12
course management systems, 211
craft and design, 165–71
Creative Commons, 49, 125, 135, 139, 157, 196, 228–29
creative content, as domain of UGC, 147
creative production, regulation of, 220
criminal infringement, 101, 102, 110–13; similarity of to secondary infringement, 112
Criterion, 87, 194, 256n10
criticism, fair dealing and, 73, 74, 77, 151, 163, 206, 211
critique, vs. parody, 163
Crocker, Amanda, 231
Crown, ownership by, 95–96; copyright term for, 50
Crown copyright, 96, 237; abolition of, 248n2
Cuisenaire v. South West Imports, 46
cultural knowledge: appropriation of, 220; Indigenous, 220
cultural nationalism, 31
cultural property, Indigenous, 220–24
Cultural Property Import and Export Act, 247n7
cummings, e.e., 52

damages, 105, 107–8, 249n3; limits to, 107–8; statutory, 107–8
databases, idea-expression dichotomy and, 47
data sets, 146; government, 147
Davies, Gillian, 22; *Copyright and the Public Interest*, 23
Death by Popcorn, 154
deep linking, 144
Defoe, Daniel, 18
delivery up, 105
Delrina v. Triolet Systems (2002), 46
Department of Canadian Heritage, 29
derivative rights, 55
design: craft and, 165–71; definition of, 166; fifty-object limit, 166–68; graphic, 165; industrial, 166–67
device prohibitions, 115

Dickens, Charles, 21
Diderot, Denis, 22
digital archives, 100, 201, 232
digital copying, 190
digital locks, 15, 31, 113–17, 131; circumvention of, 29, 113–17, 204–5; control types, 114; exceptions to, 115–16; prohibition types, 115; rights management information and, 116
digital material, pricing and contract terms for in LAMs, 212–14
digital media, 134–48; limitations of, 142–43; potential of, 142
digital music, 128–33, 185
digital resources, licensing of, 194, 213
digital rights, 173
digital rights management (DRM), 138–43, 213, 236
digital sharing, 203
digital technology, 12, 220, 232; development of, 29; LAMs and, 200, 205–7; limitations of, 142; music and, 128–33; public goods quality of, 14. *See also* digital media
digitization: of archives, 236; of journalistic heritage, 176; of library collections, 100, 212; of licensed materials, 192, 212
Dinesen, Isak, 52
direct linking, 143–44
Direct Response Television Collective (DRTVC), 88
distance education, 192
distribution, 103–4; of films, 153; of music, 129; rights of, 66
DJ work, 127, 247n4
Doctorow, Cory, 89
documentary films/filmmakers, 151, 155–56, 191, 232
Documentary Organization of Canada (DOC), 155
Donaldson v. Becket (1774), 6, 19, 240n7
downloading: definition of, 132; legality of, 29, 128; lobbying to prevent, 130; unauthorized, 130–31; vs. streaming, 132, 135
dramatic works, 37, 38; adaptation of, 63–64; commercial performance of, 112; sole rights in, 56
Dr. Bonham's Case, 5

droit d'auteur, 22–24; vs. copyright, 241n21
due diligence, 99–100
Duggan, Gordon, 242n36
DVDs: backups of, 205; copying from, 138; of televised concerts, 98

e-books, 142, 214; public library boycott of Random House, 213–14
economic analysis of copyright, 8–10
economic (section 3) rights, 54–67, 94–95, 97, 98, 151, 227
editing: of music, 129; software, 149
education, 182–98; copyright exceptions, 187–94; distance/online, 192; fair dealing and, 73, 74, 77, 178, 183, 184–87, 188, 192–94, 195, 197, 201, 206, 211, 235–36, 247n5; funding for, 232; K-12, 185, 235; public support for, 230
educational institutions: definition of, 188, 245n5, 255n7; elementary, 81, 193; post-secondary, 183, 193, 195–98, 226, 235–36; secondary, 81, 193
employment, copyright while in, 94–95
encryption, 135; encryption research, 83, 85, 115; of software, 141
England: Acts of Parliament, 5–6. *See also* United Kingdom
engravings, 39, 68
Entertainment Software Association (ESA), 136
e-reserves, 207–9
errors and omissions (E&O), 153
ESA v. SOCAN (2012), 62, 64, 132, 136, 210
exams, reproduction for, 189
exclusion mechanisms, 11, 13–15, 141, 223; hybrid nature of, 14; legal, 141; for movies, 143; technological, 14, 141
exclusive rights, 20; copyright holders and, 4; vs. non-exclusive rights, 174
exhibition: of films, 153; public, 64; standard rates for, 159
exhibition rights, 64, 158–60, 199
expression. *See* idea-expression dichotomy
Exxon/Mobil, 143

fabric, patterned, 166
Facebook, 13, 147
fair dealing, 59, 60, 71–73, 98, 100, 103, 117, 127, 151, 153, 155, 177, 234, 236, 254n8(ch.14); accepted purposes for, 73; and alternatives to dealing, 79–80, 82, 187; and artists, 162–64; and character of dealing, 78–79; definition of, 75; e-books and, 142, 213; education and, 73, 74, 77, 178, 183, 184–87, 188, 192–94, 195, 197, 201, 209, 235; and effect of dealing, 80, 187; on Internet, 135, 137–38; iTunes music clips and, 160; and journalism, 172, 180, 181; and libraries, 200–203, 205–6, 209; and nature of original, 80; photography and, 151; and purpose of use, 77–78, 81; statutory basis of, 73–74; six fairness factors for, 77–80, 151, 184–85, 186, 203, 259n2; substantiality requirement and, 74, 79, 81; tests for, 76–80; and UGC, 148; vs. fair use, 74–75
fair use, 74–75, 185, 201–3, 208–9, 249n3
fairy tales, 6
fan fiction, 146
fan vids, 146
fashion design, 253n4(ch.13)
Faulkner, William, 52
Federal Court of Appeal, 41, 81, 88, 128, 160
Feist Publications v. Rural Telephone Service, 240n10
Fewer, David, 242n35
fictional work: idea-expression dichotomy in, 46; sole rights in, 56
file-sharing: legality of, 29; monetization of, 89
filmmakers: amateur, 149, 153; documentary, 155–56; independent, 157; student and emerging, 153
filmmaking industry, 150; view of Canadian law by, 112
films, 38, 149–57, 191; commercial vs. non-commercial use of footage, 157; exclusion mechanisms for, 143; sole rights in, 56; stills from, 178
first ownership, 94–96; rules governing, 95
first publication rights, 62
Fisher, William: "Theories of Intellectual Property," 7, 238n2
fixation, 42–45, 47, 179, 191
Flickr, 137, 147, 149
folk copyright, 168
Foreign Reprints Act (1847), 25

formality requirement, 36–37, 47–49, 244n16
Forsberg, Walter, 157
found-object art, 162
Fournier, Constance, 58
Fournier, Mark, 58
framing, 144–45
France: copyright law, 22–23; copyright legacy of, 22–24; National Convention, 22; perpetual moral rights in, 68
free culture, 228, 230
freedom of speech/expression, 21, 219, 223
freelancers, 95–96; journalists, 173–76; and publishers, 176
free/libre software, 147, 228
free market system, 8
French Declaration of the Rights of Man, 4
Friedland Report, 248n8
Frye, Northrop, 43
Fung, Richard, 156

galleries, 64, 159, 160–61; funding for, 232
Gaylor, Brett, 155
Geist, Michael, 30–31; *From "Radical Extremism" to "Balanced Copyright,"* 139
General Agreement on Tariffs and Trade (GATT), 32
General Agreement on Trade in Services (GATS), 32
general infringement. *See* infringement, primary
General Public License (GPL). *See* GNU General Public License
Georgia State University, 208. *See also Cambridge University Press v. Becker*
Germain, Doric: *La Vengeance de l'orignal* (The Vengeance of the Moose), 70
Ghandl (Haida storykeeper), 222
"gift" economies, 225
Glenn Gould Estate v. Stoddart (1998), 179
The Globe and Mail, 137–38, 154, 175, 176, 178
GNU General Public License (GPL), 227–28
Gone with the Wind, 163, 253n4(ch.12)
Google, 212, 232
Gould, Glenn, 179
government funding, 230–33
government publications, 96
Grand Upright v. Watner (1991), 127

granting councils, 233
graphic design, 165
graphics on articles, 166
Guo, Edward, 53

Hager v. ECW Press (1999), 245n3
Haida culture, 222
Halbert, Debora, 148
Harper, Stephen, 30
Harvard College v. Canada, 243n2
HathiTrust, 201, 251n5
Hayhurst, William, 37
Hermesh, Michael: *The Baggage Handler*, 109
Hesse, Carla, 21
Hesse, Hermann, 52
Holden, John, 232
Holmes, Oliver Wendell, 21
honour: of author, 68–70, 126, 160; of clan, culture or nation, 221
hot market information, 12
human expression, nature and characteristics of, 11
human rights, 4
hypertext links, 143–45

idea-expression dichotomy, 42, 45–47
ideas, nature and characteristics of, 11
Imperial Copyright Act (1842), 24–25
implied consent, 135, 137
improvisational music, as citation economy, 225
incentivization, 220
incidental inclusion exception, 151–52
indictment, 112
Indigenous cultural knowledge, 220
Indigenous cultural property, 220–24, 226, 228, 230
Indigenous Peoples Council on Biocolonialism, 260n3
Indigenous protocols, 223, 228
Indigenous traditional knowledge (TK), 220–21; in public domain, 221
Indigenous vs. non-Indigenous approaches, 222
individual rights, 4, 6
industrial design, 166–68; registration of, 167–68
Industrial Design Act (1985), 166–67, 243n1
industrial-scale production, 166

infomercials, 88

information: economics of, 141; as information, 12; nature and characteristics of, 11

information goods, 8, 11, 141, 230

information society, 11

infringement, 30, 47, 71–74, 125, 128, 137, 141, 144–45, 151–52, 154, 164, 171, 172, 177, 181, 183, 188, 191, 201, 205, 208–10, 218, 224; civil, 102–10; civil vs. criminal, 101–2; commercial vs. non-commercial, 107–8; conviction vs. indictment for, 112; criminal, 101, 102, 110–13; defence of, 103; digital locks and, 113–17; exceptions to, 83–86, 115–16; on Internet, 137; and knowledge, 104–5, 112, 116; moral rights, 68, 160–62; plagiarism vs., 224–25; prevention of, 142; primary, 102–3, 105; secondary, 102, 103–5, 112; steps to follow in cases of, 106–7; substantiality requirement for, 79; trademark, 150, 177; unintentional secondary, 105; vs. users' rights, 112–13; website linking and, 62

injunctions, 105

Instagram, 149

installations, 162

insurance companies, E&O and, 153

intangible vs. tangible goods, 10–11

Integrated Circuit Topography Act (1990), 243n1

integrity, right of, 68

intellectual creations, as public goods, 10–15

intellectual goods: and exclusion mechanism, 14–15, 141; as non-rival in consumption, 14, 141

intellectual labour, 5

intellectual production, 4

intellectual property (IP), 5, 15, 32; alternatives to, 219; "big four" of, 35–36; law, 220. *See also* TRIPS, World Intellectual Property Organization

intellectual property rights (IPR), 20, 223; expansion of, 8

interlibrary loans (ILL), 205–7

International Convention for the Protection of Performers, Producers of Phonograms, and Broadcasting Organizations, 250n6

International Federation of Library Associations and Institutions (IFLA), 237

International Federation of the Phonographic Industry (IFPI), 241n31

International Intellectual Property Alliance (IIPA), 236

International Music Score Library Project, 53, 251n5

international treaties and trade deals, 236–37

Internet, 134–38, 149, 220, 230; cartoons from, 140; circulation on, 64, 65; copyright infringement on, 137; distribution of teaching materials via, 183; footage on, 157; libraries and, 212; linking from, 143–45; music and, 128–33; publicly available material on, 190, 193–94; reusing work on, 137; visual arts and, 170

Internet Service Providers (ISPs), 30, 64, 89, 131

Internet trolls, 109

interoperability, 83, 85, 115

interviews, 179

Inuit artists, 164, 220

iPhone, 147

iPods, proposed tax on, 130, 242n33

Ireland, 241n24

iTunes, 15, 83, 89, 133, 138, 160, 187, 235; and paracopyright, 142

Jefferson, Thomas, 12

jewellery, as art or design, 169

Johnson, Samuel, 8

joint authorship, 39, 224; copyright term for, 50

jokes, 44

journalism, 172–81; literary, 176

Joyce, Sarah, 242n36

Judge, Elizabeth, 96

Jusquan (Haida writer), 222

Kane, Paul, 49

Kay, Jonathan, 58–59

Kennedy, John, 241n31

Kirstaeng v. John Wiley (2013), 246n11

Knopf, Howard, 242n35

knowledge: criminal infringement and, 112; digital locks and, 116; nature and characteristics of, 11; primary infringement and, 105; secondary infringement and, 104

knowledge-based economy, 8

knowledge sharing: and knowledge building, 227; public interest in, 30
known author, copyright term for, 50
Koons, Jeff, 164

labour theory, 238n2
Lacan, Jacques, 244n12
Lamb, Brian, 89
LAMs (libraries, archives, and museums), 200. *See also under* archives; libraries; museums
later viewing, recording for, 83, 84
law enforcement, 115
Law Society of Upper Canada (LSUC), 72, 210
layering, 146
Le Chapelier, Isaac, 23
lectures, 39, 44–45; reporting on, 85
Lefsetz, Bob, 129
legal citations, guide to, 262–64
legal clinics, 110, 111
legal reform, 237
Lessig, Lawrence, 228–29
lessons: definition of, 191–92; telecommunication of, 190, 191–92
L'Estrange, Sir Roger, 18
Leuthold, Catherine, 109–10
liability, fear of, 226
Liberal Party, 29, 31
libertarianism, 230
librarians, 93, 186, 200–201
libraries, 199–214; copyright advocacy and, 143–46; copyright exceptions for, 203–5; cost recovery and revenue generation in, 211–12; and fair dealing, 200–203; and funding, 232; management and maintenance of collections of, 203–5; patron services in, 210–12; photocopy machines in, 67, 200, 210; pricing and contract terms for digital materials in, 212–14; special collections in, 211
Libre Expression (Free Expression), 70
licence fees: educational institutions and, 183, 226; for Internet material, 30
licensing, 194–98, 235; blanket, 88, 194; collective, 131, 183, 194; general, 194; individual vendor, 194; libraries and, 212; of music, 125; preservation and continued access, 212–14; temporary, 176; terms of, 135;

transactional, 88; vs. tariffs, 258n21
linking, 143–45; deep, 144; direct, 143–44; framing, 144–45
Linux, 138
literary journalism, 176
literary works, 37, 38, 46; adaptation of, 63–64; sole rights in, 56; translation of, 63
Litman, Jessica, 43
live performance, in educational institutions, 189
Locke, John: *Second Treatise of Government*, 4–6
Loewenstein, Joseph, 18
Logorama, 155
logos, 150–51; for 2010 Vancouver Olympics, 222
Lovell, John, 25
lyricists, 122, 124

MacLaren, Eli: *Dominion and Agency*, 27
McPherson, Isaac, 12
magazines, covers of, 178, 181
Magna Carta, 4
Markie, Biz, 127
Massachusetts Institute of Technology (MIT), 229, 232
Mattel, Inc. v. 3894207 Canada Inc., 243n2
McDonald's, 151, 155
McGill, Meredith, 21
McKeon, Michael: *The Secret History of Domesticity*, 18
McLachlin, Chief Justice Beverley, 42
McLean, Wallace, 52
McMahon, Kevin, 153
McOrmond, Russell, 242n35
mechanical rights, 126
media, digital. *See* digital media
media corporations, demand for rights by, 177–78
media ownership, concentration of, 172
Mexico, copyright term in, 49
Michalak, Sarah, 214
Michelin v. CAW (1996), 71, 163
Milliken & Co. v. Interface Flooring Systems (Canada) Inc., 249n1(ch.8)
Milton, John, 3
Ministry of Canadian Heritage, 261n11
Mitchell, Joni, 156

Mitchell, Margaret, 253n4(ch.12)
mobile digital devices, 149
mods, 147
Moiseiwitsch, Carel, 177
Moldaver, Justice Michael, 136
monopoly, copyright as sort of, 55, 219
Monsanto Canada Inc. v. Schmeiser, 243n2
Montgomery, Lucy Maude, 49
Moodie, Susanna, 25
moral rights, 54, 67–70, 95, 97, 98, 125, 159,
    160–62, 168, 221, 237; performers', 218; of
    silent performers, 246n12; waiving of, 97,
    122
Mount, Nick: *When Canadian Literature Moved
    to New York*, 27
Mukurtu CMS, 228
Murray, Gordon, 177
Murray, Laura, 242n35
museums, 199–214; cost recovery and revenue
    generation in, 211–12; funding for, 232;
    management and maintenance of collections
    of, 203–5; patron services in, 210–12; special
    collections in, 211
music, 65, 121–33, 149; licensing of, 125; online
    previews of, 132, 187, 219; private copying
    of, 86, 91
musical works, 37, 38, 66, 122; adaptation
    of, 63–64; commercial performance of,
    112; copyright term of, 124; definition of,
    249n2(ch.9); in online video games, 136; sole
    rights in, 56
Music Canada. *See* Canadian Recording Indus-
    try Association
Music Copyright Infringement Resource, 127
music download services, 15
music rights: twentieth-century, 122–27;
    twenty-first-century, 128–33

Nadel, Ira, 96
Napster, 131
National Film Board (NFB), 157, 232
National Gallery of Canada, 160
National Post, 245n3
natural law, 4, 6, 8, 9, 22; approach to property,
    4–5; constraints of, 5
natural rights, 22
neighbouring rights, 40

Neighbouring Rights Copyright Collective
    (NRCC). *See* Re:Sound
network effects, 13
new business models, 236
news broadcasts, recorded, 189, 191
newspapers, 25, 173–78, 205–6
news reporting: fair dealing and, 73, 74, 77, 151,
    181, 206, 211; ownership of copyright in,
    173–78
New York Stock Exchange, 143–44
non-traditional subject matter, 39, 40
North American Free Trade Agreement
    (NAFTA), 32
North American Free Trade Implementation
    Act (1993), 241n31
Nozick, Robert, 7
NPD, 249n1(ch.9)

online learning, 192
online resources, 236
Ontario Association of School Boards, 258n26
Open Access educational resources, 194, 214,
    228–29, 232–33, 236
Open Courseware (MIT), 229, 232
open source: knowledge, 227–30; materials,
    196, 220; software, 147
oral presentations, 44–45
Orbison, Roy: "Oh Pretty Woman," 163
original composers, rights of, 122, 125
originality, 40–42, 243n7; criticism of, 43
"orphan" works, 99–100, 131, 237
O'Sullivan, Gilbert, 127
ownership: corporate, 224; by Crown, 95–96;
    determining, 93–100; first, 94–96; Indigen-
    ous, 222; registration and presumption of,
    48; unlocatable, 99–100, 140
owners' rights, 29, 31, 54–70, 150, 178;
    enforcement of, 101–17; and e-books, 142;
    exceptions to, 83–86; on Internet, 137; justi-
    fication for, 5; limitations on, 5
Oxford University Press, 208

P2P (peer to peer), 131
Page, Steven, 130
paintings, 68; reproduction of, 161
Panitch, Judith, 214
Papin, Wade, 169

Paradis, Christian, 107

paracopyright, 142

Parker, George L.: *The Beginnings of the Book Trade in Canada*, 26

parody, 162–64; fair dealing and, 73, 74, 77, 151, 177, 206, 211, 247n5; law permitting, 31

password protection, 135–37, 141

pastiche, 163; law permitting, 31

patent, 11, 20, 35, 36, 243n2; vs. copyright, 244n16

Patent Act (1985), 239n12

patronage, 4, 17

patron services, in LAMs, 210–12

pay-or-destroy requirement, 191

PDF files, 137

perceptual difficulties, persons with, 85, 115–16

performance rights. *See* public performance rights

performances, 39; at agricultural fairs, 86; definition of, 61, 123; in educational institutions, 189–91; of recordings or broadcasts, 189, 191; religious, educational or charitable, 86; right to fix, 125

performers' performances, 37, 39, 45, 61, 98, 179, 191; copyright term for, 50, 124, 126; definition of, 123–24; first ownership of, 95; moral rights in, 69; rights in, 55

performing arts organizations, funding for, 232

periodicals, small, 254n8(ch.14)

permissions, 224; cost of, 156–57; minimization of, 153–55

perpetual copyright, 5–6

personal copies, 83, 84, 115

personality theory, 238n2

photocopying, 57, 66–67, 184; in educational institutions, 88, 183; self-serve in libraries, 67, 200, 210

photographs/photography/photographers, 39, 51, 138, 149–57, 179; amateur, 149–50, 153; art, 156; locating ownership of, 99–100; professional, 150; reproduction fees, 157; student and emerging, 153; technological protections for, 135; use of commissioned, 86

physical resources: appropriation of, 5; as perpetual, 5

piracy: consumer vs. corporate, 177; industrial-scale, 143; music, 133; prevention of, 112

plagiarism: as insult to honour and legitimacy, 225; vs. copyright infringement, 224–25

Plant Breeders' Rights Act (1990), 243n1

plates, 39

political speeches, reporting on, 86

positive law, 6

posthumously published works, copyright term for, 52

post-secondary institutions, 183, 235

PowerPoint presentations, 140

precedent, legal, 5, 95, 239n8, 250n8

premises, definition of in educational institutions, 256n8

price discrimination, 142

price system, 14

Princeton University Press v. Michigan Document Services, 208, 247n2(ch.5)

printers, lobbying for licensing scheme by, 25

printing enterprises, rise of in U.K., 18–19

printing press, 17

Prise de Parole v. Guérin (1996), 68, 70

Prism Hospital Software v. Hospital Medical Records Institute (1994), 63

privacy, invasion of, 143

Privacy Act, 247n7

private goods, vs. public goods, 11, 13

private music copying, 86

private property, copyright as form of, 49

private study, fair dealing and, 74, 77, 185–86, 205–6, 211

product placement fees, 150

Professional Writers Association of Canada (PWAC), 104, 174

programming, 146, 227

Project Gutenberg, 251n5

property, 222; cultural, 220–24; intellectual, 5, 11, 32, 35–36, 222; literary, 22; natural law approach to, 4–5; personal, 11; products of mind as, 22; tangible vs. intellectual, 5

property rights, 5, 8, 9, 22, 223

proprietary images, 150

public art exception, 152

public broadcasting, funding for, 232

public communication rights, 64

public domain, 49–53, 124, 127, 131, 138, 142, 177–78, 211, 221, 224, 232, 251n5

Public Domain Day, 52
public exhibition, 64
public funding, 230–33
public goods: intellectual creations as, 10–15, 141; vs. private goods, 11, 13, 141
public interest, 176
Public Lending Right, 233
Public Library of Science (PLOS), 229
publicly available material (PAM), 190, 193–94, 251n1
public performance, 146, 159, 190–91, 256n10; criminal offence, 112; rights, 61, 125, 199, 245n7
public reading, 86
public resources, private appropriation of, 4–5
public shaming, 254n5(ch.13)
publishers: and educational market, 183; monopoly of, 17–19; music, 122; newspaper/magazine, 176–77; rights of, 19; small, 231, 254n8(ch.14); v. consumers, 134
publishing industry, hampered development of, 27–28
Pyrrha Design Inc., 169

Queen's University, 182
quotations, substantiality requirement and, 60

radio, 65, 90; apparatuses, 115; college, 90–91; stations, 88, 125
Randall, Alice: *The Wind Done Gone*, 163
Random House, 213–14
Rankin, Matthew, 154
Rauschenberg, Robert, 162
Raustiala, Kal: "Fashion Victims," 253n4(ch.13)
Readex, 251n5
reasonableness standard, 218
record companies, 122
recording industry, 122–33
Recording Industry of America, 249n1(ch.9)
record labels, 122, 124, 129, 130
Redwire, 222
Reel Injun, 155
Regroupement des artistes en arts visuels du Québec (RAAV), 160, 164
Reid, D.C., 91
remix, 146, 164; non-commercial, 218, 236
rental, rights of, 65–66, 245n7

reprints, American, 24–27
reproduction: alternatives to, 143; of broadcast, 84; computer use and, 245n2; downloading as, 64; education and, 189, 193; as infringement, 137; libraries and, 199, 203–4; music and, 129; owners' rights and, 137; of paintings, 161–62; of photographs, 107, 157; for private purposes, 84; substantiality in, 137, 151; temporary, 85; users' rights and, 178
Reproduction of Federal Law Order, 248n1
reproduction rights, 55–60, 64, 70, 72, 125–26, 135, 159, 199, 245n7
reprography, 88
reputation: of author, 68–70, 126, 160–61; of clan, culture or nation, 221
resale rights, 164, 237
research: academic, 225–26; fair dealing and, 74, 77, 185–86, 201, 205–6, 211; public support for, 230
reserve materials, 207–9
reserved rights, 142
Re:Sound, 125
Re:Sound v. Motion Picture Theatre Associations of Canada (2012), 133
responsibility, Indigenous ownership and, 222
revenue generation, in LAMs, 211–12
review, fair dealing and, 73, 74, 77, 206, 211
Richardson, Henry, 239n11
rights: adaptation, 245n7; assigned, 97–99; association, 68; attribution, 68, 126, 168; authorization, 66–67; authors', 6–7, 19, 21, 22, 63, 67–70, 94, 238n1; booksellers', 19; broadcast, 125; communication, 199; consumers', 30; conversion, 63–64, 245n7; derivative, 55; digital, 173; distribution, 66; economic (section 3), 54–67, 94–95, 97, 98, 151, 227; exclusive, 4, 20; exclusive vs. non-exclusive, 174; exhibition, 64, 158–60, 199; first publication, 62; of fixing performances, 125; human, 4; individual, 4, 6; integrity, 67; of journalists, 172–81; licensing of, 177; mechanical, 126; media corporations', 177–78; moral, 54, 67–70, 95, 97, 98, 122, 125, 159, 160–62, 168, 218, 221, 237, 246n12; music, 122–33; natural, 22; neighbouring, 40, 122, 124; in non-traditional subject matter, 40; original composers', 122, 125;

owners', 5, 29, 31, 54–70, 101–17, 137, 142, 150, 178, 235; of owners of sound recordings, 126; performance, 125, 199; property, 5, 8, 9, 22, 223; public communication, 64; public performance, 61, 245n7; publishers', 19; rental, 65–66, 245n7; reproduction, 55–60, 64, 125–26, 159, 199, 245n7; resale, 164, 237; reserved, 142; to restrain publication, 173; as restrictions for someone else, 10; as severable and freely assignable, 55; sole, 55, 56, 159; telecommunication, 245n 7; transfer of, 93, 159, 173, 176; translation, 63, 245n7; users', 29, 66, 71–86, 89, 112–13, 125, 135, 138, 150–53, 163, 178–81, 183, 184, 188, 212, 214, 224, 235–36; waiving of, 97, 122; web publication, 176; in works and other subject matter, 55

rights-based theories, 4–7

rights clearance, 88–89, 98–99, 126

rights management information (RMI), 116, 140

ringtones, mobile phone, 88

Riopelle, Jean-Paul: *La Joute*, 252n3

RiP! A Remix Manifesto, 155

rivalry in consumption, 11–14, 141

Robertson, Heather, 175

Robertson, Kirsty, 169–70

Robertson v. Thomson (2006), 40, 175, 176, 254n2, 254n8(ch.14)

Rogers, Art, 164

Rogers v. Koons (1992), 164

Rogers v. SOCAN (2012), 132

Rombauer, Irma, 52

Roosevelt, Eleanor, 52

Roots, 151

Rose, Mark, 18, 43, 238n1

royalty payments, 164

Sackville-West, Vita, 52

Sage Publications, 208

Saint-Amour, Paul, 43

sampling, 127, 129

satire, 162–64; fair dealing and, 73, 74, 77, 151, 206, 211, 247n5; of social attitudes, 164

Saturday Night, 181

Saussure, Ferdinand de, 244n12

scanners, 57, 210

scholarly journals, subscriptions to, 213–14

school boards, 183, 226

schools, funding and, 232

Schwarzenegger, Arnold, 30

science, public support for, 230

sculptures, 39, 68

Sea in the Blood, 156

security: copying for purposes of, 85; national, 115

separability requirement, 166–67

"serials crisis," 214

service prohibitions, 115

sewing patterns, 171

Shameless: The ART of Disability, 155

share-alike licence, 229

Skaii (Haida storykeeper), 222

small claims court, 110

small-press publications, 176

small-scale tools, as domain of UGC, 147

Smith, Roberta, 162

Snow, Michael, 70

Snow v. Eaton Centre (1982), 68, 70, 161

Snow White movies, 6

SOCAN v. Bell (2012), 73, 78, 82, 83, 132, 160, 163, 183, 185, 187, 200–201, 219

SOCAN v. Canadian Association of Internet Providers (CAIP) (2004), 64, 65, 73, 131

social planning theory, 238n2

Société civile des auteurs multimédia (SCAM), 87–88

Society for Reproduction Rights of Authors, Composers, and Publishers in Canada (SODRAC), 126–27

Society of Composers, Authors and Music Publishers of Canada (SOCAN), 64, 65, 87, 88, 124–26, 136, 235; tariffs, 88, 247n3(ch.6)

software: control of uses by, 138; encryption of, 141; free, 147, 228; open source, 147

sole rights, 55, 56, 159

Songwriters' Association of Canada, 89

Sony, 156

Sony Corp. of America v. Universal City Studios, Inc., 240n10

sound recording maker, definition of, 250n3

sound recordings, 37, 39, 66, 98, 146, 191, 203; copyright term for, 50, 124; definition of, 126, 133; first ownership of, 95; perform-

ances of, 189; rights in, 55, 56, 122; rights of owners of, 126; sampling from, 127. *See also* music
soundtracks, 133
South Shore Public Libraries, 213
special collections, 211
Spotify, 235
Standing Committee on Canadian Heritage: Interim Report on Copyright Reform (2004). *See* Bulte Report
Stationers' Company, 17–19
Status of the Artist Act (1992), 159
Statute of Anne, 6, 17, 19, 20, 37
statutory damages, 107–8
Sterbak, Jana: *Vanitas*, 243n10
storytelling, 45, 220
Stowe, Harriet Beecher: *Uncle Tom's Cabin*, 63
Stowe v. Thomas, 63
streaming, 192; definition of, 132; as telecommunication, 64; vs. downloading, 132, 135
strict liability, 105
students' work, copyright in, 193
substantiality requirement, 57–60, 98, 137, 151; and fair dealing, 74, 79, 81; five-part test for, 58–59; and sampling, 127
summary conviction, 112
Suntrust v. Houghton Mifflin, 253n4(ch.12)
Super Size Me, 155
Supreme Court of Canada, 7, 29, 40–42, 65, 67, 71, 73, 75, 77–83, 88–89, 92, 132–33, 150, 161, 163, 175, 184, 187, 193, 197, 234, 235. *See also under individual legal cases*
syndication, 173

tangible goods/property, 5; vs. intangible goods, 10–11
tariffs, 195, 235, 247n3(ch.6), 250n5; blanket, 88; proceedings, 89; "upward creep" in levels of, 92; vs. licences, 258n21
taxation, 95, 233
teachers' work, copyright in, 193
technological exclusion mechanisms, 14
technological neutrality principle, 134, 136, 207, 210
technological protection measures (TPMs), 14, 113–17, 135, 140, 193, 204–5
technology enhanced learning, 192

telecommunication: of lessons, 190, 191–92; right of, 245n7
Tele-Direct v. American Business Information (1998), 41
television, 157; time-shifting of recorded shows, 218, 236
terms of use, website, 134–38
Teva Canada Ltd. v. Pfizer Canada Ltd., 243n2
Théberge, Claude, 161
Théberge v. Galerie d'Art du Petit Champlain (2002), 57, 65–66, 71–72, 73, 161–62
Thelwell, Jane, 169
thumbnails, website, 160
tolerated use, 247n4
Toronto Star, Pages of the Past archive, 177
tort liability, 95
total market failure, 14
trademark, 35, 36, 105
Trade-marks Act, 234
trade secrets, 35, 36
traditional medicine, 220
transactional licensing, 88
transaction costs, 9
transformativity, 259n2
translation, 146; rights of, 63, 245n7
Trans-Pacific Partnership (TPP), 32, 236
treaties, 31–32, 236–37
tribute, 163
TRIPS (trade-related aspects of intellectual property), 31–32
Trudeau, Pierre, 96
Twitter, 147

United Kingdom: Act for the Encouragement of Learning, 182; censorship in, 21; copyright legacy of, 17–19; duties placed on U.S.-originated publications, 25; Licensing Act, 17–18
United States, 29, 235–37, 238n7; and Berne Convention, 28; book industry, 20; Constitution of, 7, 21; Copyright Act, 21, 27, 166, 201–2, 240n9, 244n1, 246n11, 249n3; copyright law, 74–75; copyright legacy of, 19–22; copyright term in, 49; damages, 249n3; Digital Millennium Copyright Act (DMCA), 116–17; expansionism in, 240n16; fair use, 74–75, 185, 201–3, 208–9, 249n3;

first sale doctrine, 66, 251n6; fixation, 243n8; Innovative Design Protection Act, 253n4(ch.13); lack of federal government copyright, 96; lack of moral rights provision in, 68; Patent Act, 240n9; pro bono legal clinics in, 110; reciprocal copyright arrangement with U.K., 21; refusal to grant copyright to foreign authors, 21; reprints of British books, 24–25; Semi-Conductor and Chip Protection Act (1984), 240n9; Supreme Court, 21, 163, 238n7; test for fairness, 75, 203; unlocatable ownership, 99; Visual Artists Rights Act, 68

United States Trade Representative (USTR): Office of, 112; Watch List of, 31

universities, 236–36; and collectives, 88–89; faculty of, 95, 193; and fair dealing, 82, 185; libraries, 214; and licensing, 194–96; and publicly funded knowledge, 226

University of Michigan, 201

University of Toronto, 195

University of Western Ontario, 195

unknown authors, copyright term for, 50

unlocatable ownership, 99–100, 140

unpublished works, 61; copyright term for, 51–52

uploading, 128, 135

Urban Counterfeiters blog, 254n5(ch.13)

urban spaces, trademark and copyright material in, 150

use controls, 114, 257n15

user-generated content (UGC), 83, 84, 135, 145–48; domains of, 147; non-commercial, exception for, 152–53

users' rights, 29, 66, 71–86, 125, 163, 183, 184, 224, 235–36; advocacy for, 89; clinic for, 111; education and, 183, 188; on Internet, 135, 138; journalism and, 178–81; libraries and, 212, 214; strengthening of, 183; vs. copyright infringement enforcement, 112–13

utilitarianism, 7–8, 9, 20, 22, 23, 238n2

Vancouver Olympics (2010), logo of, 222

Vancouver Sun, 177

Vaver, David, 44, 72, 79

Veuve Clicquot Ponsardin v. Boutiques Cliquot Ltée, 243n2

video. *See* films

video games, 149, 246n9; online, 136; and para-copyright, 142

video mash-up, 247n4

virtual worlds, 147

visual art, 158–64; collective action and advocacy in, 158; technological protections for, 135

von Finckenstein, Justice Konrad, 128

Warman, Richard, 58

Warman v. Fournier (2012), 58–60, 234

weather reports, 45

"web of science," 226

web publication rights, 176

websites: consent and terms of use for, 134–38; images from, 178; linking to, 61; thumbnails on, 160

Weekend, 179

Wheaton v. Peters, 238n7

Wiebe, Christle, 221

Wikipedia, 147, 212, 227

wikis, 147

Willmore, Danielle, 169

Winnipeg Free Press, 154, 178

Winnipeg Jets, 154

World Intellectual Property Organization (WIPO), 24, 250n6; Copyright Treaty (WCT), 31; Intergovernmental Committee on Intellectual Property and Genetic Resources, Traditional Knowledge and Folklore, 260n3; Internet Treaties, 29, 31; Performances and Phonograms Treaty (WPPT), 31

World Trade Organization (WTO), 24, 31–32. *See also* TRIPS

World Trade Organization Agreement Implementation Act (1994), 241n31

Writers' Union of Canada, 90, 201

writing back, vs. parody, 163

York University, 197–98, 234

Young-Ing, Greg, 223

YouTube, 147, 149, 152

Zorich, Diane, 211–12